TECHNOLOGY HOW INFLUENCES HUMAN BEHAVIOR CHANGES

JOHN LOK

Made with ♥ on the Notion Press Platform
www.notionpress.com

Contents

Foreword

Prologue

Contents

Why social behavior may influence organizational strategy needs to be changed ?
How and why human behavior may influence economic growth or recession?
Reasons why human behavior may influence economic recession or growth ?
p.31-45

Chapter 2
Labors behavior or robot improves organizational performance

Reasons organizations need to raise employee individual performance

- Reasons why a company shouldn't ignore Performance Appraisals
- Considerations When Giving Employees Raises Performance, what benefits organizations
can earn
- Using Base Wages to Give Raises
- Performance-Linked Raises Performance.
- Merit-Based Raises and Your Compensation Package Can Raise Employee Individual Performance
- Reviewing Your Compensation Package
- Five Ways to Monitor Employee Performance Can Bring Your Organization Long Time Raising Efficiency And Productivity
- Learning Employee individual emotion feeling in performance evaluation
- How do organizations need to determine pay raises for employees as well as why it has relationship to improve performance ? p.46-60

How organizational structure influences
employee individual performance

- How organizational Structure can influence organizational behavior or employee individual performance? p.61-70
- Reasons organization structure influence employee performance
- Why does Reward management (RM) need?
- APPRAISAL METHODS Management By Objective
- The Effects of Organizational Structure on Behavior p.71-80

Negative influences to lacking evaluating

Robotic how influences global economic change

● Encouraging new economic development competitive factor

● How much robotic productivity can raise competition to cause effects of limited competition disadvantage to our society.

● How and why robotic brings long time intellectual development to global factories ?

● How can robotic help manufacturers to raise productivities or improve efficiency?

● Why does production possibilities curve may explain why robotic participation may assist few workers to raise productivities in factories ?

Must Developed And Developing Countries
Need Artificial Intelligent To Replace Human Job

● How AI help developing countries to communication and agriculture and learning and medical delivery development

● Emergency Response to developing countries' earthquake natural damage sudden occurrence predicting

● Smart AI Agriculture

● Medicine Delivery to developing countries' patients urgent need

● Assistance to reduce teaching work workload or psychological pressure to teachers in developing countries' schools

● Why does smart phone help developing countries communication ?

● The Positive Impact of Mass Media in Developing Countries

● Why do developed countries need to develop AI

● AI may bring what benefits to developed countries

● How can AI be dangerous when developed countries continue to develop AI to become weapon to replace soldiers?

● Why the recent interest in AI safety ?

● Why do developed countries people need AI ? p.116-129

Reference

Artificial intelligence bank service working environment

● The Future of Artificial Intelligence In The Workplace

● A positive future with artificial intelligence to bring remote online office working environment chance

● Artificial Intelligence (AI) in Banking working environment

AI technology can bring better Customer Support in bank service environment

● What does artificial intelligence mean for the bank service office worker?

How AI influences our daily working culture in any office working environment

● The Future of Artificial Intelligence In The Workplace

● A positive future with artificial intelligence

● How AI can help office workers to do tasks more easily

● How AI is eliminating repetitive administrative tasks

● What are traditional office tools to be replaced by AI ?

How artificial intelligence can raise office efficiency

● Training presents challenges

An office effort measured approach to AI

● How can AI be dangerous to office working environment?

● Five ways to use AI to improve business efficiency to these office tasks

How can AI help offices to save energy or avoid to waste energy ?

Can AI replace office workers

● AI insurance workers

● AI public service workers

● AI replace warehouse workers

● AI can replace counter cashier service staffs

● AI can replace accountants in accountancy service industry

Why Developed And Developing Countries Need Artificial Intelligent To Assist Office Tasks

● How AI help developing countries to communication and agriculture and learning and medical delivery development

● Emergency Response to developing countries' earthquake natural damage sudden occurrence predicting

● Smart AI Agriculture

● Medicine Delivery to developing countries' patients urgent need

● Assistance to reduce teaching work workload or psychological pressure to teachers in developing countries' schools

● Why does smart phone help developing countries communication ?

● The Positive Impact of Mass Media in Developing Countries

● Why do developed countries need to develop AI

Why and how human behavior may influence the country's economic growth or recession?
Technology how impacts human behavior changing?
How and why employees behaviors may influence economy development?
Robots invention whether they can help organizations to raise efficiencies or inefficiencies?
Why social behavior may influence organizational strategy needs to be changed ?
How and why human behavior may influence economic growth or recession?

Chapter 3 Technology how helps businesses changes labor effort
Organization efficient using robotic resources economic method
- What is efficient use of resources to any organizations in economics?
- What is meant by economic using of resource to organizations?

p. 161-175

Amazon Organizational Intangible Robotic resource
Ecommerce organization resource management strategy
Why does in this e-commerce organization situation, online webstore speed must be the most important factor to influence its sale success. Information technology internet speed, online webstore design, online transation convenience transaction feeling (tangibale and intangible both resource factors) may influence its future clients number ?
How an ecommerce organization resource affects ssociety?
Why has online webstore's information technology resource close relationship to influence online buyers number and social job chance?
How do organizational resources affect organizational change?
How to build organizatonal resource using right psychology
How psychological method help employees to avoid resource waste
Can Robotic Help Warehouses To Avoid Resource Waste On Behavioral Economic View
- Behavioral economy view whether robotic can help
warehouse to avoid time and human resource
waste
- How Amazon warehouse applies logistic robots to help
it to waste resource waste

Organizational intangible resource management strategy

● How to implement effective intangible resources management strategy to achieve performance improvement to Amazon e-commerce organization? Why does non intangible resources effective negligence management to Amazon organization, it will cause worse performance to Amazon e-commerce organization any one e-buyer?

Robots whether can help organizations to avoid
resource waste
How robotic technology help organizations reduce waste? p.176-190
How to use technology to help reduce waste?
How green robots are helping with environmental sustainablity?
How robots may help organizations to avoid resource waste?
How robots are applied to reduce waste in hotel organizations?
Can Robotic Help Warehouses To Avoid Resource Waste On Behavioral Economic View
● Behavioral economy view whether robotic can help
warehouse to avoid time and human resource
waste
● How Amazon warehouse applies logistic robots to help
it to waste resource waste

Management science role solves
resources shortages
● What does management resource science mean ?
● Why does management science help organizations to avoid/solve resource shortage?
● How to apply management science technological tool to solve resource shortage to organization challenge?
● Reasons to organizations need management resource strategies ?
● How applying right science resource management principle to organizations
● How Amazon e-commerce applies resource management principle to bring avoiding resource waste benefit?

Robots whether can help organizations to avoid
resource waste
Can Robotic Help Warehouses To Avoid Resource Waste On Behavioral Economic View p.191-218
How robotic technology help organizations reduce waste?
● Behavioral economy view whether robotic can help

warehouse to avoid time and human resource
waste

- How Amazon warehouse applies logistic robots to help
it to waste resource waste

Organizational robotic intangible resource management

- How to implement effective intangible resources management strategy to achieve performance improvement to Amazon e-commerce organization? Why does non intangible resources effective negligence management to Amazon organization, it will cause worse performance to Amazon e-commerce organization any one e-buyer?

Can apply robotic to improve consumer behavior to improve performance
Why can (AI) be applied to help businesses to predict consumer behaviors ? p.219-240
How can artificial intelligent tools predict consumer behavior in vehicle market retail industry ?
What is (AI) consumer behavioral prediction tool?
How any why will (AI) tool assist manufactures to attempt to predict consumer behavior before and after consumption occurrence?
What is (AI) deep learning techniques to help farmers to raise food growth
How can apply (AI) technology to help book shops to decide book sale price
How can (AI) provide businesses with better-informed decisions?
How can artificial intelligent tools predict travelling consumer behavior in airline and air agent travelling market ?
Why is (AI) big data gathering tool better
than psychological and survey methods to predict traveler individual travel choice behavior?

How (AI) big data determine future travel behavior from past travel experience and perceptions of risk and safety for the benefits to travel consumers?

How (AI) influences organizational change
Can artificial intelligence development to tourism industry

- Artificial intelligence will bring what benefits to
Influence traveler consumption behavior or raising
Travelling leisure desire?

(AI) raises driven automation industry development

1.1 (AI) - driven automation industry development how to influence work nature change p.241-260

Competitive influences between artificial intelligence and human job to improve efficiency and performance
(AI) journalism, media publishing, digital communication technology trend

How robots help manufacturers to raise productivities

How Robots bring positive impact to raise organizational productive change

Becoming robotic talent human how influences social changes

CHAPTER ONE

Technology or human behavior whether may influence economic growth or recession

Human Behavioral network job brings social economic benefits

What does human network job mean ? Why may human network job be popular? Why human network job behavior may influence economy ? Nowadays internet is popular to use. We can apply internet to find data , search any new things, even earn money. Why does internet may become huma network job source. For example, e-publish may be one kind of new human network job. Any authors may apply internet channel to help them to sell electronic or paper books from e-publisher web store. They may apply facebook, you tub etc. any online channel to promote themselves new books to let new readers to know whether when they may buy themselves favourable new topic books to read from electronic publisher web store.

Thus, future electronic publisher industry may help any authors to build internet network platform to help them to sell and promote ot advertise their any one new electronic or paper book topic to let global any one reader to choose to buy their any new topic books from electronic publisher web store easily and conveniently. However, it implies that electronic network platform author may be one kind of future new human network job in our societies.

How electronic network platform author job may bring economy benefit in macro economy view? A person can have few friends, contacts and still be very influential if these few
friends and contacts are themselves highly influential, e.g. one author must not need to know any one reader in global society. When they like to choose any electronic books from electronic internet network platform. They may become the author's any one topic book buyer, when they feel the author's any one topic book is fun and attract they make decision to buth the strange author whose the topic book from electronic book publisher's platform web store conventiently in short time. Although, they are strangers, they do not know themselves , but the reader can understand what it way that made Google from writing platofrm to create new creative mind and typing network job method to replace traditional hand writing book method for global authors. It will be one kind of new human network writing job.

Hence, global any one reader can apply an innovative search engine , such as google.com to find whether whom author personal new topic books are value to read from internet.
Then, the electroniuc publisher's web store may be new book store platform sale network to help the author to sell many electronic or paper books from electronic network platform
in short time. So, internet may be future new network plaform to help global any one author to create network writing job absolutely. Furthermore, internet may be popular social media
to help any one author to build goold relationship between his/her readers. It is one kind of new network, human network job. New authors do not need to buy many paper books to prepare to put in any one book shop warehouse. Their every book can print on demand to reduce out of book stock in any one book shop. They may choose to sell either electronic books or paper books both from any one book publisher web store. So, electronic network platform may be one kind of good writing channel to help human authors to create income and it can also help authors to bring new creative mind and new topic fun content books to let readers to know and buy to read from electronic publisher network platform.

Why does human behavior may be one kind of new human network job to bring global economic advantages. ALthough, it may be free income or without inocme, but the person does the network behavior, his/her behavior may be bring advantages to influence many other people's health. For this case, when a worker in a coffee shop in an airport gets a vaccination aganinst the flu, it does not only helps him or her stay healthy, but also helps the many travellers who might otherwise have been inflected if that workers caught the flu. So, the externality , the result implies the vaccination of even a part of a community conveys benefits to the whole community. For example, governments pay special attention to the vaccinations of school children, teachers, health mothers, and the elderly, categories of people particularly susceptible not only to catching, but also to transmitting a disease.

It is not accidential that governments are heavily involved with vaccination . When there are externalities, free market, fail to persuade individual incentives with society's

their the worker's decision of whether to get a vaccine ends up attracting whether other people get sick. The workers might not fully take all these other people's potential suffering into account when making her or his vaccination decision.

As Stanford University does many suggestions, understand this and tries to help them make the right decisions and so providers free flu vaccines for its staff and students.

Small pockets of unvaccinated individuals can allow a disease to gain a spread more widely well-being. For example, parent weighing the costs and benefits of a vaccine for their child is not always thinking of the consequences of that vaccination to other people. THese are markets in which subsidizing or regulating behavior can make everyone better off. Because the reason for requiring that a child be vaccinated before enrolling in school is not just to protect that child, because each child's vaccination affects others via potential contagions.

Robots take our jobs behavioral and economy influences

Robot job behavior brings economy influences

If one day robots can replace human to do simple, even complex jobs. They will bring what influences to our global societial economy.The popular economic refrain declares that the

global middle class is dying and robots will soon take our jobs, e.g. shopping center customer service jobs, library service jobs, cinema ticket sale jobs, restaurant kitchen cooker jobs,

even, bus drivers, taxi drivers etc. public transport driving jobs, accountant, doctors etc. professional jobs. Whether it is beautiful or petty matter if our future societies have many human jobs can be replaced to do from robots.

Businessman must may reduce to employ employees and reduce to pay salary or wage, when robots can be replaced to do their employees tasks. But, societies must bring unemployement rate rises , due to societies will have many people loss jobs when their employers choose to buy robots to serve their clients or do any office tasks or customer service or cleaning etc. tasks.

In micro economy view, employers may save money in long term, but in macro economy view, it will cause unemployment ratio rises , even crime rate rises when there are many people lose

jobs in societies. These models of doom, though, fail to account for the hundreds of businesses riding the waves of change in their industries when robots may be invented to replace human to do many simple , even complex tasks in our future societies.

WE may image that one small factory needs to manufacture fishes canes to sell to supermarket, the small , cheaper stuff and higher margin parts of the fishes manufacture industry. Before, this factory needs to employe many human factory workers need to help every fresh customer makeing the perfect fishing gear, designed for performance, durability, and cost in order to achieve to manufacture every fish cane in whole fished processing manufacturing stages. Every worker needs to spend about 15 to twenty minutes to finish every fish cane , till to delivery to any supermarket to sell. If this fish canes manufacturing factory can apply manufacturing robots to help them to finish any one working tasks , every robot can only spend five minutes to finish whole fresh fish cane manufacturing process. Thus, every robot can

help this factory save 10 to 15 minutes time to finsh every fish cane manufacturing process. IN fact, time is money, because when every robot can help this factory to reduce 10 to 15 minutes time to compare human worker. Then, this factory can finish about 20 fish canes in one hour if it can use robot to help it to manufacture fish canes. Otherwise, if this factory still use human workers to help it to manufacture fish canes, then it can finsh about 3 to 4 fish canes in one hour. SO, the manufacturing efficiency ensures that robots must help this fish manufacturing factory to raise fish canes number more than human workers. So, in robotic behavioral economy view, manufacturing robots must help this fish canes manufacturing factory to raise fish canes manufacturing number and deliver increasing number to supermarkets to prepare to sell every day. Robots can help this fish canes manufacturing factory bring manufacturing time saving,

rising manufacturing efficiency, improving performance and reducing wages expenditure long time advantages in micro economy view. However, manufacturing robots can also bring disadvanages to society, e.g. increasing unemployment ratio, increasing crime rate,
this factory workers will lose jobs and income, they need earn social welfare from government and increasing government finance pressure in short time, even long time in macro economic view.

Stanford University graduate program in economics, Scott lecturer explained that "in demand and supply economic theory for robots supply and demand case, robots supply number increasing may influence human workers demand number decrease. It sometimes calls " the efficient frontier".
No specific human beings were mentioned in any of economics classes. As robots supply and demand in market case, They (robots) may be purely theoretical " agents" who reached to the most reasonable sale prices in order to persuade any one businessman buyer to make manufacturing robot buying decision whether robots can help him / her to bring how much saving time , saving money, saving cost, improving performance, efficiency economic benefit before he/she plans to reduce workers number when he/ she decides to apply robots to replace human workers in his/her factory or office or any service department, e.g. cinema ticket sale service, shopping center customer service, shopping center cleaning , supermarket customer service etc. service or sale tasks. When robots can replace human to do any one of these tasks in any organizations. So, robots may be human worker agents who reached to prices the way robots would react to a software
command. There was nothing that explained why some people thrived and others did n't or why truly brilliant, hardworking people could fail when much lazier folks succeeded." Having been admitted to the Stanford University graduate program in economics, Scott lecturer hoped to get his answers there.

How robots influence our future social changing? Using the right technology can be a boon to your business in this economy. For internet example, it is easier than ever to find well-matched customers all around the world, to stay in contact with them, and to more quickly design the products they want. If you focus solely on being cutting -edge, though you risk letting the technology
take over what should be very robust relationships with your customers , employees, and colleagues. IN nowaddays society, technoligical advances

and cutomation, personal

relationships in business are more crucial than ever. I mean that robots can not replace human to serve clients to let them to feel more comfortable and passion more easily. For shoe shop case example, if the shoe shop apply one robot to serve its clients to replace human shoe salesperson to serve its shoe customers. Robots ensure that they can not persuade every shoe potential buyer to make shoe buying decision more easily when robots need to contact every shoe potential buyer. The reason is simple, because robots can not touch any one shoe buyer individual emotion very easier.

If the shoe buyer needs the robots to help him/her to choose any right shoe styles when he/she can not feel himself / herself can make the most right shoe style choice decision. The robots can not replace human shoe salesperson to make shoe style choice judgement more easily. They must need longer time to analyze whether which shoe style may be the most suitable to the shoe buyer. Otherwise, human shoe salesperson may attempt to make the most right shoe style choice decision to help any one shoe buyer to chooce the most right style shoe because he/she owns shoe style sale experience, shoe style knowledge, the most important reason is that they can feel every shoe customer individual emotion to touch whether he/she will feel comfortable or happy when they attempt to help every shoe customer to seek the most right shoe style in every shoe customer whole shoe searching processing. Othwerwise, serving robots are only one machine, they can not touch or feel every shoe customer individual emotion whether he/she feel comfortable or unhappy or happy when they need to contact them in whole shoe searching processing. Hence, I believe that some tasks robots can

not repalce human staff to do very easily. Otherwise, robots may bring disadvanatges to let any one businessman to loss his/her customers, due to robots can not touch every customer

emotion to compare human staff in service tasks more easily. Robots serving customer behaviors may cause money lose and customers number lose to the shop in micro economic view.

Intellectual human economic behaviors

What does intellectual human economic behaviors mean ? I believe that when we choose or decide to do intellectual behaviors, then our societies will be influenced to bring economic growth in consequence.I shall attempt to indicate pollution case to explain how and why eithet our intellectual or foolish behaviors may bring economic growth or recession in consequence

as below:
On one hand, for air pollution social case aspect example, if we only consider to buy cars to drive for working aimr or holiday leisure aim. Then, our societies air will be polluted. Our health will be influenced to bad. Our car driving behaviors may cause global environment air pollution serously. In long tiem, global air pollution will bring our bodies health to be bad. Although, ourselves car driving behaviors may bring our driving travelling leisure enjoyment and comfortable feeling in short time, also we so not need to pay public transport fare often, but we need to compensate ourselves health economic intangible loss due to air pollution , when cars number increases, dirty air will cause ouselves health to become bad.
In the result, we will need to pay more medical expenditure when we are old age, due to ourselves bodies will become bad, due to we breathe global dirty air every day, due to ourselves cars pollute air in long time, e.g. 10 to 20 years, even 30 more without limited air pollution environment. So, driving cars behavior may be one kind of human foolish behavior and our foolish behavior may bring ourselves future long time medical expenditure absolutely.
One the other hand, water pollution social aspect, if we often keep much rubblish to pollute sea, oil exploration porcessing pollute ocean , ships gas pollute ocaen, then fishes will eat polluted food and drive dirty water, due to global ocean is polluted.
In fact, because human only to conside how to buy boats to carry on leisure enjoyment activities, or catch cruises to travel on the sea. Also, oil manufacturers only consider researching anywhere to find new oil exploration places to manufacture oil product, when their oil exploration processes pollute ocarn . Consequently, global fishes drink polluted warer or eat polluted food. They will have poison. SO, human will have high chance to eat poison polluted fishes, due to fishes are poison or are polluted. So, human is doing foolish activities, we only hope to find oil exploration places to pollute ocean or we only spend money to buy ticket to catch ships to travel anywhere in global ocean. All of these human foolish behaviors will bring pollution to global ocean. On consequently, we will need to compensate to eat polluted or dirty or poision fishes, ourselves bodies health will be bad. In long time, we need have high chance to pay medical expenditure when we are old. So, pollution case may be one good example to explain how and why human foolish behavior may influence ourselves future need to compensate serious medical loss.

All of these human foolish behavior will bring pollution to global ocean. On consequently, we will need to compensate to eat polluted or dirty or poison fished , ourselves bodies health will be bad. In long time, we will have high chance to pay medical expenditure, when we are old. So, pollution case may be one good example to explain how and why human ourselves intellectual or foolish behaviors may influence future long time economic loss or economic growth or recession in micro and micro economic view.

On another water pollution aspect hand, if we often keep rubbish to sea, oil exploration processing pollutes ocean and ships' gas pollute ocean, then fishes will eat polluted food and drink dirty water, due to fishes will eat polluted food and drink dirty sea water because the global ocean is polluted seriously.

In fact, because human only consider how to buy boats to carry on any leisure water activities, or catches cruises to travel on the sea. Also, oil manufacturers only consider any where to find oil exploratin places to manufacture oil products from ocean, when their pol exploration processes can plooute ocean. Consequently, global fishes drink polluted water or eat direty food. They will have poison. So, human will have high chance to eat poison fishes.

Otherwise, such as pollutin case, it can infuence inflation or deflation. Consequently, the reason indicates supply and demand theory. If air pollution is serious, then we will consider health issue, global cars demand number may be influenced to reduce, when global cars number demand will reduce, global car prices and supply number will need to change to fall down in order to attract or persuade global car consumers choose to make car purchase decision.

Hence, global car manufacture number and car price will be influenced to reduce, due to global air pollution issue. Consequently, deflation will occur because when the country citizen usually does not spend much extra saving money to buy car expensive goods. Money value will be low. Otherwise, if global cair pollution is not serious, human considers to buy cars to enjoy driving leisure lives. So, global car demand is influenced to increase , also global car price will also influenced to increase.

Consequently, gobal human will choose to buy cars to drive. Due to we accept to spend extra saving to buy expensive car goods. Car sale price and supply may be influenced to rise up. Money value is influenced to reduce. Inflation may be influenced, due to global car consumers number increases, we would not have extra money to spend easily. Car expensive goods expenditure influences our spending habit to avoid to make car purchase decision more easily. So, human intellectual or foolish activities may bring inflation or deflation consequency in possible indirectly in macro economic view.

On conclusion, above pollution case explain that how and why human intellectual or foolish economic behaviors may bring inflation or deflation consequency as wll as economic growth or recession consequency as well as any goods demand and supply increasing or decreasing consequency. It implies that human behavior may have indirect relationship to influence any goods demand and supply number to either increase or decrease result as well as any goods price will be influenced to increase or decrease in micro and macro economic view.

The relationship between social change and human behavior

Why does economic changes may influence human individual behavioral change? I shall attempt to indicate shopping behavior and staying at home behavior to explain their case and effect relationsip as below:

Human behavior can be influenced by economic change or economic change can be influenced by human behavior? Why does recession may influence consumers reduce shopping desire? In social recession suitation, it is possible that many people lose jobs suddenly, due to businessmen lose many customers. They need to make decision to reduce employees number in order to continue to keep businesses. Consequently, many firms (organizations) their employees may lose jobs. When they have much time, due to lose jobs, they will feel to avoid to spend too much time and money to go to shopping often. Many losing jobs people, they will often stay at homes. So, they will reduce time to go to shopping, then non essential products won't their preferable choice purchase products. Hence, recession will change many losing jobs people their shopping or consumption desires to avoid to buy non essential products often . Usually when economic boom, many people have jobs to do because consumers number must increase when many people have jobs to do. Then, many people can accept to spend money to buy non essential products often. Many people feel spend time to go to shopping can satisfy their purchase of any kinds of new products useful psychology or desire. So, recession is one good example to explain it can influence many people do not like often to leave homes to go to shopping easily. Many people like to stay at homes, becaue they feel worry about spending too much shopping time when they leave homes. Their staying home time is one good negative shopping behavior example. So,

economic change may influence human individual behavior changes , they have direct cause and efect relationship in behavioral economic view.
May human behavior influence economic change? Is it possible that human behavior may bring the country social economic change in macro economic or micro behavioral economic view ? I shall indicate publishing industry example. Do you feel that if there are many students feel learning is very important when they read many books or many of students feel interesting to read or they have reading new books in habit, then it is possible that the country will have many students like to spend time to go to any book shops to choose the books, they feel that they can help they learn new knowledge. Then the country will increase students number, they often spend time to visit any one book shop every week. Their visiting book shops behavior which may become their habits. So, the country will increase students number, they often spend time to visit book shops. Also, it implies that visiting book shops behaviors may be their behavioral habits.
So, when the country has many students often spend time to visit book shops , their visiting book shops behaviors may help any one book shop to raise books sale chance. So, the country's student individual often visiting book shop behaviors, their habitual visiting book shops behaviors must may assist help any one book shop to increase books sale number absolutely.
Consequently, any one book shop , its books sale bumber must be influenced to increase to increase because the country will have many students like or feel need visit book shops habit in order to choose any suitable books to buy to read at home in order to raise themselves learning effort. When the country has many bok shops often have many students visit their book shops, then their books sale number may be influenced to increase. It explain why student individual visiting book shop behavior may help any one book shop sale number increases also.

How human productive behavior may influence economic development

May any country which citizen behavior assist themselves country development? It is one cause and effect economic question. I mean that if the country itself citicen can not concentrate mind or energy to choose to do one kind of industry in order to let themselves country can bring the most benefit, then whether the counry itself economy can bring the most serious economic benefit. I shall attempt to indicate these countries themselves indistry choice to explain whether these countries themselves citizen productive behavior may help themselves countries to achieve the largest economic benefits. I shall indicate as below:

New Zealand farmer individual wine productive behavior

For New Zealand country example, this country concerns itself effort is foucs on farming agricultural aspect. So, this country has many farmers concentrate on farming agricultural aspect. May New Zealanders choose to spend time to produce different kinds of wines, e.g. wine or red grape wine is for the people are eating meat, or they are eating dinner.

When these New Zealanders their behaviors choose to do farming or agriculture to grow and produce different kinds of taste of white or red grape wine drinking products job. Themselves grape agriculture behavior will influence these New Zealanders themselves, they can learn how to improve different kinds of grape wine drinking products in order to achieve every kinds of white or read grape wines taste improving aim during their white or red grape producing process.

Why can New Zealander every individual white or read grape wine producers improve their white or read grape wine taste more easily? In behavioral economic view, it can explain that why any one New Zealander white or read grape wine producer can be encouraged or excited or persuaded to concentrate nervous and energy and effort to learn how to improve their white or red grape wine products easily.

In fact, New Zealand is one agricultural food export country. It has good natural environment resource , e.g. land, seed to provide any one farmer to produce themselves any kinds of agricultrual food products, e.g. fruit, or wine food products. Because New Zealanders know themselves country has enough natural resource . So, in common, many New Zealanders choose to attempt to do farming agricultural jobs in order to export themselves any kinds of fruit or meat or wine products to overseas or sell to domestic in order to earn profit.

So, when these New Zealand farmers number has been increasing every year. This country farmers will feel themsleves competition between this New Zealand farmers themselves are serious due to they may feel New Zealanders choose to do agriculture businesses in order to export themselves different kinds of farming food to overseas or sell to local to earn profit.

Hence, when many New Zealand farmers feel that farmers number has been increasing every year. They will feel themselves competition is serious. They must need to spend much time and nervous and effort to research what method is the best how to produce the best taste of white or red grape wine products in order to let local or overseas wine buyers to choose to buy

his/her producing white or read grpae products to drink.
Hence, in competition psychological view, may influence many New Zealand white or reaad wine producers had been beginning to change their learning behavior on researching what method is the best in order to produce the best quality of taste red or white wine products to sell in order to attract overseas or local white or read grape wine drinkers to choose to buy his/her wine products. Their behavior will focus on learning how to raising or improving white or read grape wine taste method more than only focus on producing a large number white or red grape wine products. They believe wine quality is more important to compare wine producing number. So, New Zealand wine producers themselves wine producers behaviors have been changing on concentrating on researching wine quality method aspect more then wine producing number aspect in behavioral economic view.

America high technological productive behavior

For America example, US is one high technological country, it owns many high technological knowledge talent inventors, e.g. computer science inventors. Hence, US must attract many diferent countries owning high technological computer inventors choose to go to US to develop their computer science profession career. Also, it seems that when many computer science inventors or professions choose to go to US to develop themselves computer science new career. In behavioral economic view, due to their leaving themselves countries choice, which may bring influence themselve country job behaviors need to be changed. They must need to adapt US new live. Because they will forgive their past computer science job. These computer science professionals need to spend time to adapt US new lives. They " past computer science job behaviors" will need to be changed to their new US any computer employer's new computer science job model.
Because their traditional computer science jobs needed to be forgot in their themselves countries. They will feel their old computer science job knowledge and behavior needed to change in order to let their US any one new of computer company employer feels satisfactory to accept their new working behavior in any one US computer organization.
So, on the other hand, many US computer company employer will feel that they must need time to accept any one new overseas computer science professions their working behaviors, their working attitude daily, because these foreign comouter science professional, their past computer working

behaviors and working attitude must be different to US domestic computer science professions.
In behavioral economic view, these overseas computer science professions, their working behaviors and attitude must be needed to change in order to adapt any one US new computer company itself domestic or local computer science professional stafs themselves daily working behaviors and attitude because these overseas and local computer science professionals must need to team work together.
In behavioral economic view, it is only one way that foreign computer science professionals must need to change themselves past country traditiona daily working behaviors and attitude in order to cooperate with these US local computer science professionals in teams more easily.
Consequently, if these foreign compute science professionals can change their past working behaviors and attitude to let any one US local computer science professional feels to cooperate with them easily in short time. Then, the US computer company itself whole computer professional teams themselves efficiencies will be influenced to raised or improved by the changing past working attitude and working behaviors of these foreign computer science professionals. So, in behavioral economic view, only if US any one computer company hopes itself computer teams themselves efficiency can be raised or improved when it decides to employ foreign computer science professionals and US domestic computer science professionals. They need to work in teams together. They must need to let these foreign computer science professionals to know how to change their working behaviors and attitude to let their domestic computer science professionals feel easy to work together. Then, the US computer company itself whole team efficiency must be rasied or improved easily in short time.

● China share market investing behavior

For China share market example, economic development depends on financial market. Because if many Chinese have interest to invest to carry on shares buying and selling activities in orde to learn how to earn shares interest and share profit when the China shareholder can make decision to sell himself/herself shares in the the high price, then he/she can earn money when he/she can sell the China company's shares in the high sale share price position.
If China has many Chinese like to spend time to carry on investing shares activities. Themselves shares buying and selling behaviors will influence China has many companies can increase fund from many Chinese

shareholders in order to have enough money to expand or develop themselves businesses in China in long term.
Consequently, when China can have many Chinese like to attempt to carry on buying and selling shares investing behaviors in China share market. Themselves buying and selling shares behaviors can help many Chinese companies have effort to increase enough money or capital in order to continue to do their businesses in long term absolutely. So, it explains why when many Chinese become shareholders , they can assist China will have many companies continue to develop their businesses if many Chinese like to carry on shares buying and selling investing behaviors in long time in China financial investment market nowadays in behavioral economic view.

Why has any individual country have many people invest share behavior which can influence the country's macro consumption desire?
I shall apply shares market buying and selling investment behavior to explaiin why shares investment behavior which may impact the country's overal consumption desire as below:
In behavioral economic view, I assume that when the coutry has many people have interest to attempt to carry on shares buying and selling investment behavior, then their frequent shares buying and selling behaviors which may bring negactive consumption desire or shopping desire of these shares investors their consumer behavior.
The reason is simple, when the country has many share buyers number suddenly been increasing rapidly. Consequently, these large group share investors must need to spend much time to research any kinds of company shares variations, whether when their share prices will rise up of fall down in order to achieve buying the company's shares in the lowest price and selling the company's shares in the highest price level in order to earn profit.
Basic on this reason, they must need to spend much extra time to research share prices changing behavior every day, e.g. one working person will wait to leave his/her job, after he/she can spend time to gather data to research the day's share price changing behavior after dinner. So, the working person's right time may be his/her share price market research behavior. Before he/she may spend his/her night time to go to shopping after dinner, but nowadays, he/she will fogive to do his/her shopping behavior before dinner or after dinner at hight sometime. He/she will make decision to spend much night time to turn on computer to click on share market website to research his/her share purchase choice to investigate whether

his/her share price whether it rises up or falls down at the moment in order to make his/her share buying or selling decision at ever night time.
I mean the when the country has many people are share investors, their shares investment behavioral spenging time which will influence many shops lose customers at might often because the country will have many people feel need to spend night time to turn on computer or watch television to investigate share price variation. So, the country will have many people / share investors choose to stay at home in order to carry on share price variation investigation behavior, they need to listen share market update news from radios or watch the share market update news from computer or TV at home every night. Consequenly, they must reduce times to leave themselves homes at night. So, their shopping behavior also will be reduced. Because these share investors feel need to spend time to investigate share price variation news at homes which can bring economic benefits (high opportunity benefits) when they choose to forgive to leave homes to go to shopping times (opportunity cost) every night.
On conclusion, it seems that when the country has many people are share investors, then their share price investigating behavior may bring negative shopping emotion at night. Consequently, the country's any one shop may lose many customers from this share investor consumer group in behavioral economic view. Hence, when the country's share investors number had been increasing rapidly, it will influence any shops lose many customers from this share investing customer group at night frequenly in short time, even long time in behavioral economic view, because their shopping desires or shopping emotion will be brought negative feeling when they make decisions to spend much time to listen radios or watch TV or computers share price update nes at night. Hence, share market will bring negative impact to influence consumer shopping desire or negative shopping emotion in behavioral economic view.

Can technology influence human shopping behavioral change?
Nowadays, technological development has reached mature stage, whether technological mature stage may bring positive or negative shopping emotion influence to global consumers. I shall aplly internet inventin or ecommerce shopping channel tool to explain whether internet technology can bring postive or negative influence to global consumer behavior in behavioral economic view.
Internet is a good technological tool, it brings e-commerce business chance.

In fact, commonly, global has have many businessmen choose to use internet channel to carry on their products transactions between global online-buyers and their electronic websites. So, global many shoppers had begun to feel online shopping is more convenient to compare visiting shops shopping. Their shopping behaviors have been changed from internet technological tool. Global has many shoppers choose to buy any products from any overseas or local businessmen their web stores. They only need to spend time to find any businessmen their webstores to choose the most suitable products to pay visa to buy from their webstores. at homes. So, in general, global had have may shoppers had changed their shopping behaviors from visiting shops to visiting webstores at homes often.

So, it seems that internet technological tool had influenced global many shops disappear, but internet webstores will be replaced their actual shops on streets. Some of businessmen either they choose webstores to replace shops or choose websotes and shops both or still keep shops only. Hence, internet tool influences global businessmen have three kinds of products sale channels to let globa local and overseas consumers to choose how to buy their products.

However, in fact, many of global shoppers, youngers and olders had begun to accept to buy any products from webstores. They feel to spend time to leave homes to visit shops , their shopping behaviors will be wasted time to not essential part to their daily lives. Hence, since internet technological invention, it had changed many consumers their traditional visiting shops shopping habit to change to buying products from webstores channel.

However, on the one hand, internet creates webstores ecommerce shopping channel to let global many consumers do not need to leave homes to go to shopping. It brings negative visiting shops shopping emotion to global general consumers nowadays. But on the other hand, it also brings positive visiting internet webstores shopping emotion to global general consumer nowadays. So, it seems that global many consumers feel that they often do not need to spend much time to go out shopping. Many global consumers feel convenient and enjoy to choose any products to buy from different internet webstores, when the online buyer chooses the most suitable product, he she only needs to pay visa card to buy the product from the online seller's webstore conveniently at home.

Hence, online shopping can bring economic benefit to online buyers, e.g. avoiding walking time or spending transport fare to visit the shop to go to shopping, shortening or reducing shopping time to do another important matter.

On conclusion, global many consumers began feel online shopping can bring more economic benefits on shortening shopping time, avoiding transport fare spending aspect. So, online shopping will be popular shopping behavior for future long time. It may encourage global many shoppers can make rapid shopping decision in short time in order to carry on any products buying transaction to global any one online shopper in short time easily in behavioral economic view. So, global many businessmen had begun to build themselves one attraction webstore in order to persuade different countries consumers to choose to click themselves webstores from internet channel to buy any kinds of products in short time easily.

So, internet technology had changed consumers traditional shopping behaviors to build positive online shopping emotion as well as raise online sellers' any products sale chance easily in behavioral economic view.

Why and how human behavior may influence the country's economic growth or recession?

When one country has many people choose to do the same matter for one period, whether their behavior may influence the country's pvera; economic growth or recession . I shall attempt to indicate cases toexplain their relationship as below:

For flowing rubblish behavioral case example, do you feel that when the country has many people often flow rubblish on the streets, instead of their flowing rubblish behavior may bring streets dirty? But, their flowing rubblish behavior may explain that this country has people may have enough money to buy food to ear, or enough cloths to wear, enough bottles of water to drink, even they may have enough money to buy new television, radio, refrigeraters , washing machines, desktops or laptops electronic home products from old to new to use in order to satisfy their living needs. So, when they flow old electronic home products, their flowing old home electronic products behaviors may seem that they have enough money to buy other new home electronic products to replace old home electronic products to use at homes.

However, it seems thaat this country ought have many people have jobs to do. So, many of them, they can easy to make purchase decison to flow any old home electronic products and buy any new home electronic products to use . Because this country has many people have jobs to do. So, they can often not use old home electonic products to become rubblishs to flow on streets after they had bought any kinds of new home electronic homes.

In fact, it also implies that this country's economy grows rapidly. So, many businesses can glow up rapdly. When they expanded their businesses, they must need to increase employees number in order to let they help themselves to raise productivity or serve their clients absolutely. So, when the country has many businesses can grow up, it seems that its economy must be better or it is improved to compare past. Due to many different kinds of home electronic products had been often bought to use by this country people in this period. So, this country's any streets can be observed that expensive electronic home products were flowed on streets anywhere. then, this country will have many electronic home products sellers can sell their home electronic products very easily. When this country has many people can find any kinds of jobs to do easily. So, due to unemploymen rate had been decreasing.

In behavioral economic view, as this many electronic home products rubblish country case, we can observe this country may have many people have jobs to do. So, consumption number has been increased long time. So, cheap food, or expensive home electronic products may be rubblish on any streets. This country's people , their flowing rubblish behaviors may be explained that many of people have enough jobs to do, so they have ability to buy any good taste food to eat or buy any kinds of expensive electronic home products to use. So, this country's economy may be improved for this long period. So, in behavioral economic view, when this country can have many electronic home products rubblishs are flowed on anywherer in streets frequently. It seems that this country will have many people have jobs to do, so it causes they often change old home electronic products or replaced them easily, when they have enough income to spend to buy any kinds of new home electronic products to use at homes easily. Moreover, their flowing old electronic home products behaviors also indicate that this country has many people their salaries may be increased in possible from their emplyers. When this country can have many different kinds of home electornic products are sold. It means that this country's electronic home products needs or demand had been increasing, due to many people have jobs to do and income increases to excite their living of needs also improve. Consequently, this country may seem have better economic improvement. We can observe from this country's electronic home products rubblish increasing income in theis period.

On conclusion, this country ought experience economic growth at this period. So, " flowing expensive electronic home rubblish increasing number

" may seem that this country's economic growth is rapidly in this period, due to many people have jobs to do as well as salaries increase in this period.

Technology how impacts human behavior changing?
Technology how influences human behavior to bring changing? For example, online share purchase and sale transaction from smart phone brings share investor can do share buying or selling transation in any where and any time conveniently, non manual driving auto vehicle, bring car owner feels comfortable and spends free time to do other matter, e.g. reading, listening mucis in himself or herself car freely. electrical energy vehicle can help car owner to reduce air polluton and it can brings the drivers do not feel drive long time in any journeys in order to avoid air pollution for environmental protection responsible car drivers in our societies. Thus, they will drive long time in any journeys when they can drive electronic energy cars to replace oil energy cars.
However, online technology can also bring consumers can choose to stay at homes to buy any things from seller individual online webstore conveniently. Such as online technology can bring shoppers do not need to spend much time to visit shops to buy any things. They can choose any kinds of products from any online sellers individual online webstores conveniently at homes. Online technology excite busy consumers can make purchase decision easily as well as it can help online sellers sell any kinds of products from internet easily.
In behavioral economic view, technology can change human behavior to be improved, it can let human feels comfortable, more free time ro use, rapid making any decisions, such as apply smart phones to make share purchase or sale transaction decision, online shopping decision, even travelling any where decision in short time, when the traveller finds the most cheap hotel accommodation room price and air ticket price frm any travel agent online tourism webstore, then the potential travel customer can follow the online hotel accommodation price and air ticket price data to make decision when to buy the air ticket from the airline travel agent or make decision when to prebook which hotel accommodation room to go to the country to travel from online travel agent tourism webstores. So, technology can encourage global any country travelers to make anywhere to trvel rapidly. If the traveler can find the country's general hotel rooms and airline tickets prices had been decreasing more sightly. The traveler may make travel decision to choose the country to travel in short time, then he/she can

prebook the country;s any hotel room and airline ticket to pay by visa fraom the country's any hotel and airline travel agent webstores., before one week, even one month or more easily. Hence, online technology can also encourage traveler individual frequent travel times to be increased, due to global travelers can find any hotel rooms and airline tickets prices from internet conveniently at homes. They do not need to spend time to visit any airline travel agent to enquire travel choice country's hotel rooms prices and airline ticket prices. They can compare global travel of countries choices ' all hotels rooms and airline agents air tickets prices to make prebook airline seat and hotel room decision before one week, one month even six months early.

On conclusion, online technology can encourage global travelers can make travelling any where and when traveling time desicions easily. It can excite tourism industry develops in long time. Also, such as electricity cars invention can encourage environment protection car owners do car purchase decision easily, because they can choose to drive electronic energy cars to replace oil energy cars in order to avoid air pollution occurs easily. So, electronic cars can increase electronic car purchasrs number, due to many of environmental protection attitude of car owners can choose to drive electricity cars to bring air cleans, even non -manual driving cars can encourage lazy driving and free time driving car owners to choose to buy non-manual (artificial intelligent) cars to drive , because they can spend much free time to read, listen music or do any matters in themselves cars, they do not need to drive cars, robotic (AI) auto driving machine is such one non-manual driver to help them to drive themselves cars confidently. So, non-manual driving cars can attract lazy and enjoying free time driving car owners to choose to buy to replace traditional manual cars to drive easily. Moreover, online share transaction can help any share investors to make share buying and selling decision in short time easily. When they can apply smart phones technological tool to carry on share buying and selling activities easily. They can observe any share rising or falling price suitation from smart phones in any where any any time easily. So, smart phone technology can help global any shareholders to make share purchase and sale transaction easily. So, technology can encourage human makes decision in short time rapidly.

How and why employees behaviors may influence economy development?

In behavioral economy view,I believe the country's any organizational employees behavior may bring indirect relationship to influence the country's long term economic development. I shall indicate past manufacture industry social development period to explain their relationship. For many countries' past business activities had belonged to manufacturing industry, such as US, UK past before 1980 year, it focused on steel manufacturing and steel manufacturing related machine products. So, US, Uk developed countries manufacturing industries may be past main country's economic income sources. I assume US , UK past had one million number different kinds of industries. They ought had about seven houndred thousand number organizational businesses were belonged to manufactured industry. They may include:

Steel manufacturing and steel related machine manufacturing, e.g. vehicle manufacturing, home appliances, e.g. washing machine, television, radio, refrigerate cooler, heater, air condition etc. different kinds of different kinds of steel -related manufacturing machine, they were manufactured from US, UK steel machine manufacturers. So, US, Uk the other three hundred thousand number industry may be general service industry, e.g. hotel service, restaurent, cinema, public transport service, tourism lesiure , wine bar, supermarket etc. different kinds of non-manufacturing industries business organizations were operated in UK, US past before 1980 year.

So, in UK, US developed countries industry development history, they ought have high percentage of businesses belonged to steel related manufacturing machine and steel products. Also, in the past before 1980 year, US, Uk business employers , they employed many workers are manufacturing workers. They needed to spend long time to work in factories. They were skillful workers, and they are trained to manufacturing cars, washing machine, television, heater, etc. even steel itself different kinds of steel related products to prepare to deliver to their shops to sell to US, Uk local or overseas clients.

So, I believe that past UK, US ought employ many employees, they belonged to skillful manufacturing workers, manufacture increasing steel machine or steel related machine number of products rapidly daily. So, if UK, US had had many of these manufacturing factories owned high skillful workers, then their manufacturing steel-related machine or steel both kinds of products number must be influenced to raise rapidly. Consequently, their steel machine manufacturing products would been exported to overseas or would been sold to local both markets , they may be influenced to raise sale number. They (these manufacturing workers) needed to be trained to know how to manufactur these different kinds of machine products in the efficient teams and they ought to be trained to raise their efficiencies in order to shorten time to manufacturing many kinds of steel related manufacturing machine or steel itself products rapidly. So , if their efficiencies and manufacturing performance was improved, these US, UK any one manufacturing worker and their teams ought achieve raising productivities significantly.

Hence, when past UK, US manufacturing industry development period, if these two countries‘ any manufacturing factories could have many manufacturing workers could be trained to be skillful and proficient manufacturing workers. Then, in past every day to these factories workers, they ought help their steel or steel related manufacturing employers to raise any kinds of machine or steel products number in every team. So, when past in the manufacturing industry development, US, UK could have many factories' manufacturing workers themselves steel or steel related machine products manufacturing skill could be trained to to improve to any kinds of these machine or steel manufacuring products quality as well as their products number could be influenced to raise by themselves skillful improvement significantly every day.

Then, what would be influenced to occur to past UK, US manufacturing industry period? In behavioral economic view, when these two manufacturing industry developed countries, such as UK, US , if they had many factories workers can be trained to improve their skill in order to achieve any kinds of steel or steel-related machine products quality could be improved as well as products manufacturing number could be also increased absolutely.

In consequence, past UK and US both countries ought increase themselves any kinds of steel and steel related machine products number to be supplied to themselves local shops to let local clients to choose any one kind of machine manufacturing products to buy easily as well as they could also export to supply overseas any countries to buy their different kinds of steel or steel related machine products to let overseas steel or steel related manufacturing machine product buyers, they can have many of these different kinds of these steel or steel-related different kinds of manufacturing machine from UK and UK these both countries easily to compare other countries.

On conclusion, I believe that past US, and UK macro manufacturing industry income GDP would increase significantly. So, they would have good economic growth performance because when many of these manufacturing workers themselves manufacturing effort could be improved. So, it explained when employees manufacturing abilities can influence economic growth indirectly.

Robots invention whether they can help organizations to raise efficiencies or inefficiencies?

In behavioral economic view, in any organizations, when the organization hopes its worker teams can raise efficiencies , the organization may choose to increase more workers number and/or it can provide training to improve these workets themselves skills in order to raise their efficiencies. For one warehouse example, when the warehouse increases many goods , they are needed to delivered these goods from the shelves to the delivering destination locations. If this warehouse supervisors feel these workers themselves goods delivery speeds are slow, which is possible due to this warehouse's workers number is not enough. So, this warehouse supervisor ought increase workers number in order to increase their goods delivery speed in order to deliver goods from the shelves to every indicated goods delivery destination in order to let any one lorry driver can transport the right kinds of goods and ensure the accurate goods number to transport to any one client home rapidly.

However, if this warehouse supervisor planed to buy several warehouse goods delivery robots to assist these warehouse workers to find the right kinds of goods from shelves and then deliver to the right destination location in the warehouse. So, these warehouse orkers can concentrate on counting the accurate goods number and ensuring the right kinds of goods in order to prepare to let lorry drivers to transport these goods to these goods of buyers themselvers homes rapidly. Consequently, in the first step, robots can concentrate on finding th right goods from shelves and delivers them to the right goods transportation of location destination. Then, in the second step, these warehouse workers can concentrate on counting the accurate goods number and ensuring the right kinds of goods in order to prepare to put them to the lorry. Consequently, when warehouse robots and warehouse workers can cooperate to work together, the most important, robots, can deal on finding the right kinds of goods and deal on delivering the accurate number of goods of job duty as well as these warehouse workers can only concentrte on counting the right kinds of goods number in order to avoid it has none any mistake of wrong kinds of goods and inaccurate goods of delivery number to be transported to the lorry and to deliver to any one buyer's home.

So, it seems that warehouse robots ought help any one warehouse worker to raise himself efficiency and avoid goods delivery of mistake occurrence easily as well as their help to warehouse workers that can let any one goods buyer feels their goods can be delivered to their homes rapidly. Moreover, warehouse robots can also help these warehouse workers to raise efficiencies because warehouse robots can help them to shorten goods delivery time between any one shelf and any one goods delivery destination of location in the warehuse because robots may help them to find the right kinds of goods from the right shelf in the short time. So, any one worker does not need to spend long time to seek anywhere is the right shelf location for the kind of goods when the kind of goods are needed to deliver to the buyer's home from lorry. Warehouse robots can help them to do this aspect of " finding the goods from the right shelf in short time job duty". So, any one warehouse worker only needed tospend less time to do the counting of any right kind of goods number and ensuring the right kind of goods job duty. Consequently, this warehouse 's any one worker, his any one kind of goods delivery time may be reduced, because robots' assistance and they may have more confidence to avoid mistake to deliver the wrong number of goods and/or the wrong kind of goods to any one goods buyer's home.

On conclusion, it seems that warehouse robots ought may help any one warehouse worker to raise efficiency for any one team in the warehouse as well as the warehouse any one supervisor does not need to spend much time to observe any one worker individual performance for " goods delivery job duty aspect" because their goods delivery job duty that had been replaced to do by these several warehouse robots. Robots can achieve the more accurate of right kinds of goods and the right number of goods delviery job performance to compare any one of human warehouse worker themselves right kinds of goods of delivery and right number of goods of delivery job performance. So, when robots can participate to cooperate with this warehouse's any one worker to do their goods of delivery job duty in this warehouse every day. Then, robots can raies any one of supervisor individual confidence in order to let they do not need to spend time to observe any one of worker individual whose goods of delivery job performane. They can concentrate on supervising any one worker whose goods transport to lorry in the final step in order to avoid to deliver wrong goods number and / or wrong kind of goods to any one goods buyer's home every day. Consequently, this warehouse's overall teams of their delviery of goods performance many be improved by robotss' participatin to goods of delivery task as well as this warehouse's oveall teams themselves efficiencies may be influenced to raise by robots' goods of delivery task participation.

Why social behavior may influence organizational strategy needs to be changed ?

Why any organizations need to know whether nowadays social behaivor how has been changing in order to implement the kind of the most right strategy to achieve the profit aim pursue in possible. I shall indicate nowadays ecommerce or online, customer shopping behavior to explain above question concerns they ought have close relationship between social behavior and organizational strategic choice or organizational behavioral changing need.

On nowadays ecommerce business, or online shopping model, this kind of shopping model in global many young and old age consumers like to apply internet tool to choose any country sellers website stores in order to stay at home to buy any kinds of products from themselves webstores in global societies.

In fact, online shopping model had been popular for long time above to twenty years. Most of global sellers will make decision to design themselves webstores in order to attract global many online buyers to choose to buy their products from themselves webstores. So, it seems that social consumers purchase behaviors had been changed to online shopping from internet invention.

Hence, social consumers purchase behavioral changes may influence any organizations' strategies need to be changed from visiting shops purchase strategy model to online purchase strategy model, if the seller still concentrate on concentrate on considerate how to design itelf , but neglects to considerate how to design itself webstore, e.g. how to design attract product photos to put on itself webstore, how to arrange sale price information location to be putted on webstore and visa card payment location on itself webstore in order to let any one online buyer can feel very easier to buy itself any kinds of products from itself webstore. Then, its potential online buyers will be influenced to increase number when they can find this online seller itself any kinds of products photes and every kinds of product sale price information and visa card payment channel locations easily from itself webstore.

So, it implies that nowadays any one seller ought need to design one webstore to let any one online overseas and domestic consumers can have chance to click itself webstore to choose any one kind of product to buy conveniently when he/she does not hope to leave him/her home to go to shop, because nowadays social shopping behaviors had been influenced to change when internet invention, them it gives another online purchase method to replace visiting shops purchase method to global any one buyer in nowadays societies.

So, if nowadays any one seller still concentrate on how to design itself shop display in order to put any kinds of product on shelf in order to let any one visiting shop customer to find the kind of product to buy, but it neglects to change to choose to pursue another new technological shopping method, such as webstore purchase method in order to implement effective strategy to design the most right webstore as well as in order to attract global overseas and local consumers to find itself webstore easily from website and find its any one kind of product phots and sale price and visa card payment button in order to choose to buy itself any kinds of products in the short time. Consequently I believe that the seller will lose many customers from overseas and local when its other same or similar product sellers choose to design themselves webstores in order to let global any one product buyer can buy themselves any one kind of product when they can pay visa card to buy their products from them webstores conveniently when they stay at home habitly. Then, the seller will lose many global potential customers in long time.

On conclusion, in behavioral economic view, any consumer behavioral social changing, which will influence any in order to avoid customers number loses significantly . In future time, organizations need to make rapid decision in order to implement the most reasonable and the most useful strategy in order to avoid global potential customers number reduces or lose them in long time. So, social behavioral changing environment ought influence any global organizations need to decide how to change themselves strategies in order to avoid customers loses significantly in future time.

How and why human behavior may influence economic growth or recession?

May ourselves daily behaviors influence our global societial continue economic growth or recession? Do they have cause and effect close relationship between human behaviors and global economic growth or recession? I shall apply behavioral economic theory to analyze and explain whether ourselves daily behaviors and our global societial economic growth or recession which have close cause and effect relationship as below:

Every country itself economic development must depend on any business activities, otherwise, any kinds of business activities must need ourselves business activities or behaviors in order to achieve any business activities as well as achieve the country's overall economic development in macro view. However, any country's overall business activites or behaviors which must depend on any kinds of individual businessmen, themselves employees daily working behavior or activity or performance in order to help them to attract or increase many clients number to acieve " earning profit" aim. So, it seems that any individual business, itself overall every department individual working behavior is one main factor to influence the company's overall business performance.

For agricultural fruit and meat food farming industry example, such as New Zealand is a farming main target industry country. It had had many New Zealanders were daily themselves own farming businesses for many years. Their farming businesses include growing fruit, sheep, cow, pig pork, meat etc. food sale business. If the New Zealand farmer owned a large size farming land, then he will choose either growing fruit or feeding sheeps, pigs, cows to be meat to to transport to New Zealand supermarkets to help them to sell to their farmers meet to New Zealanders in order to earn profit. Thus, if the New Zealand farmer owned large size of farming lands, then he needs to employ many farming employees (farming workers) to help him to carry on farming business daily tasks, e.g. picking up friuts, feeding pigs, cows, sheeps to eat food daily. These daily farming jobs are very important to influence this New Zealand farmer's meats or fruits sale number whether they can be easy or diffcult to sell in New Zealand supermarkets , if these farming workers can own encough farming knowledge or skill to know how to pick up fruits method and make judgement to know whether it is right time to pick up the kind of fruits from the trees , as well as know how feed this pigs, sheeps, cows to eat food in order to let they are better health. Consequently, their farming behaviors which can let these animals can provide the best taste and enough meat from these animals to let New Zealander to buy to eat from New Zealand any one supermarket. Even these New Zealand farming workers can know whether the kinds of fruits, e.g. oranges, apples, gapes etc. fruits whether they ought be picked up from the trees at the right time. Consequently, they can make judgement to decide to pick up any kinds of the best taste fruits to let any one New Zealander to buy to eat from any one supermarket in New Zealand. Otherwise, if they do not make judegement to know whether the kind of fruit ought not be picked up because they still need longer time to continue grow up to increase fruit size and better taste from the trees in order to let any one fruit buyer can feel better taste when they eat this kind of fruit later. If they can buy this kind of fruit to eat later, then this New Zealand farmer's his fruit buyers can buy the best taste of this kind of fruit to eat from an yone supermarket in New Zealand. Consequently, many New Zealand supermarkets will choose to buy any kinds of fruits from this farmer fruit supplier when they feel this farmer's fruits can provide more better taste fruits to compare other farmers‘ fruits.

Thus, due to New Zealand is one farming main income source country. It's any kinds of fruits and meats need to be export to overseas to sell , instead of local sale. It's GDP percent is very high to whole country 's overall income source. So, any one New Zealand farmer individual and any one farming worker individual working behavior will influence its economy whether it is influenced to grow or recession possible. Moreover, it also seems that farming workers' farming knowledge and skill will influence themselves farming daily activities to achieve the aim of the number of increase or decrease to any kinds of fruits whether they are better taste or the number of increase of decrease to any kinds of meats whether they are better taste to supply to any one New Zealand fruit or meat buyers to eat from any one New Zealand supermarket. So, it implies that any one New Zealand farming worker individual farming behavior may influence any kinds of fruits or any kinds of meat taste because they are transported to any one supermarket to sell in New Zealand.

Consequently, if New Zealans had many farmers can teach god farming knowledge and skill to let their any one farming workers know how to decide judgement to decide when it is right time to pick up any kinds of fruits from trees , or how to grow them on soil in order to let they can grow rapidly. Then, many different kinds of fruits can be provided to let any one New Zealanders can eat the best taste of fruits when their fruits are supplied to any one New Zealand supermarkets. Even, if they knew how to feed foods to pigs, cows, sheeps to eat daily. Then they can be more health and they can provide the best taste of meats to let any one New Zealanders can buy their meats from any one New Zealand supermarkets. Moreover, their fruits and meats can be transported to overseas to let any one country fruits or meats buyers can choose any kinds of New Zealand meats and fruits to buy to eat from themselves countries supermarkets. Then, many overseas fruit and meat buyers will perfer to choose New Zealand any kinds of fruits or meats to buy to compare other countries fruits or meats to buy when they go to any one local supermarkets.

On conclusion, it seems that New Zealand farming workers themselves farming behavior may influence their farming employers any kinds of fruits or meats sale number and income because their farming task behaviors must influence whether their fruits or meats taste are the better taste or worse taste to compare their other local farmers (the farmer competitors) whose fruits or meats taste. If tthe farmer's any one farming worker can be trained to learn how to know to feed animals skill and when is the most right time to pick up any kinds of fruits from trees or how to grow them on the soil methods. Due to these farming worker individual farming behavior may influence his different finds of fruits and meats sale number to be increase or decrease, so these any one New Zealand farmer must need to depend on any one farming worker whose farming working methods, if their farming working behaviors can be the best to influence any kinds of fruits to grow rapid or any kinds of pigs, cows, sheeps animals grow up rapidly , then their sale number may be increase significantly and their taste can be improved to let any New Zealand or overseas meat or fruit buyer to buy to eat to feel from any one New Zealand or overseas supermarkets, then New Zealand's agriculture industry must be influenced to increase. In the world, any one fruit or meat buyer must choose to buy New Zealand's fruit and meat to eat in prefer to compare other countries‘ fruits and meats. So, New Zealand's GDP may be influenced to raise from any one New Zealand farming worker individual farming working behaviors.

Reasons why human behavior may influence economic recession or growth?

Can ourselves daily behaviors or activies influence ourselves countries' economic growth or recession? I shall attempt to explain the reasons why they have direct or indirect relationship between human behavior and economy growth or recession as below:

I shall indicate environment pollition case to attempt to explain above question. Our societies had been experiencing servious environment pollution challenge. However, environment pollution , such as air pollution is caused by air planes and vehicles emission by air planes and vehicles emission as well as water pollution is caused by plastic rubblish, or dirty water or oil or gas chemical material, these both kinds of pollution ought may bring economic recession and this both kinds of pollution are caused by human ourselves daily foolish activities.

I believe human behavior and economy and pollution which have cause and effect relationship. I shall analyze this environment pollution case to explain why they have case and effect relationship between human foolish behavior and environment pollution and economic recession as below:

When global societies had many people like to buy cars to drive to bring emission to fresh air on the roads as well as many manufacturing factories will bring emission to pollute fresh air in their manufacturing processes. Factories and cars will bring air pollution , due to factories need to pollute fresh air in order to manufacture many products and car owners need to drive their cars to go to offices or leisure places. Their cars will also bring emisson to pollute fresh air. On consequence, car owners themselves frequent driving behaviors and factory workers themselves frequent manufacturing behaviors may bring environment pollution. Technology or human behavior whether may influence economic growth or recession. Moreover, air planes also brings emission to pollute air when they are flying in sky. Also, when ships bring oil pollution or sea plastic rubblishs bring pollution to global oceans.

In fact, manufactuers and cars owners, such as factories workers manufacturing behaviours ans car owners driving behaviors and pilots driving air planes flying behaviors and ships transport behaviors, which may cause plastic rubblish, oil or gas emission to sky or sea or on the road to cause ocean and air pollution is serious. However, human ourselves need to buy cars to drive to satisfy ourselves driving leisure or enjoyment, travelers need to catch air planes to travel to enjoy leisure needs, factories workers need help factories to manufacture many products to sell to customers to satisfy their using needs. oil exploration needs to find lands to explore new oil lands.

All of these business and leisure activites may bring serious air and water pollution. However, due to serious air and water pollution will bring earth warming challenge , such as some countries temperature will be influences to rise up to 40 degree or higher br earth warming. However, earth warming is caused by air and ocean pollution. Pollution must be caused by human ourselves, driving cars leisure and factories manufacturing business activities. Hence, if human decided to continue to do these foolish behaviors, we only pursue to manufacture different kinds of industrial products or drive cars to enjoy leisure aims, but we also neglect ourselves behaviors may bring environment pollution. Then, earth warming or earth temperature will be influenced to rise up absolutely in long term. Moreover, if our future earth will be influenced to bring serious high temperature effect by human ourselves these foolish behaviors.

On consequencey, warth warming will bring serious economic losses in possible because when ourselves earth temperature had been influenced to rise up to 40 degree or high. Ourselves health will be caused poor, due to we will feel difficult breath, we must need often tried and hard to work, due to our nervous and health will be influenced to poor by pollution and earth warming effect. Also, we need to pay more money to see doctors when we had long life. Then, our societies will lose may strong labors to help manufacturers to work, e.g. factories will reduce workers number to help manufacturers to produce more different kinds of products, due to workers health is general poor. Due to lacking enough workers to manufacture products, our societies will begin to reduce enough supply number of products to sell to global consumers to satisfy their use needs.

On conclusion, in behaviroal economic view, our societies will lose many labors due to their bodies are not health by air and water pollution. Global economic and business activities will be influenced to worse by global workers reducing number reason. So, economic recession will begin to occur in possible when pollution reaches the serious level.

CHAPTER TWO

Labors behavior or robot improves organizational performance

Amazon Knowledge innovation strategy

There are both kinds of knowledge innovation influences to Amazon ecommerce organization develops in success, one is e-commerce technological knowlegde as well as another is social innovation knowledge. I shall explain the reasons why these both knowledge may cause Amazon e-commerce organization grows in success as below:

E-commerce technological knowledge innovation

What does knowledge innovation mean? Why and how does Amazon ecommerce organization need to continue to innovate its online shopping knowledge sale channel? Can knowledge innovation impact positive online shoppers number grow to Amazon? Why does Amazon ecommerce organization apply knowledge innovation strategy to grow itself organization to achieve global online buyers number in success? In fact, when we can prepare to innovate our knowledge to be perfect. Consequently, it can help our social development more success, e.g. non manual driving vehicle, e-commerce, construction houses on water skill, hospital new robocit surgeon equipment, manufacturing robotic technology , even space touriusm leisure development etc. different kinds of technology innovation, they are depended on how human can attempt to learn to innovate or improve ourselves traditional old knowledge to be changed to any kinds of new or not discovered unique knowledge. So, human ought need to continue to learn how to innovate our old traditional knowledger to be more perfect. The question concerns how human can continue to change our old knowledge to innovate in succeed. I shall attempt to indicate how Amazon ecommerce organization can apply knowledge management innovation strategy to improve itself online puchase and sale platform to attract global many online buyers number choose its any kinds of product online purchase channel in preference as below:

What does knowledge innovation mean? The role of knowledge innovation means that knowledge management assists in building competencies required in the innovation process. Though knowledge accessibility and knowledge flow to organization, staff memebers are able to increase their skills levels and knowledge both formally and informally. An increase in skills can improve the quality of innovation as well as in society. When the country have many people can attempt to learn how to change and knowledge old knowledge to new knowledge, then they my bring their societies to develop more rapidly. So, it seems that knowledge innovation may help society, organization and individual to bring new knowledge to attribute to our future societies easily. It wil be a very important factor influence human future development in success.

However, knowlege innovation concept is not just represented by introducing or implementing new ideas or methods. The definition of meaning of innovation can be defined as a process, but involves multiple activities to uncover new ways to do things. Innovating helps developing original concepts and is to driver of optimizing operations. The purpose of innovation is to come up with new ideas and technologies that increase productivity and generate greater output with the same input on organizational aspect.

So, knowledge innovation may be applied on organization aspect, such as Amazon ecommerce organization, even individual and social aspects. On Amazon ecommerce organizational knowledge innovation aspect, innovation secures tomorrow's revenue, lowers costs and differentiates companies from the market. However, a good business model that only provides a brief market advantage and disappears after a year is not the right approach. Hence, organizational knowledge innovation aims to help the organization to raise competitive effort in long term.

Does knowledge provide innovation in Amazon ecommerce organization? Knowledge managment creates a culture conductive to tacit knowledge creation, sharing ideas in the organization, which plays an important role in the innovation process. Is knowledge management ncecessary for innovation? Beside the financial basis knowledge is the most important resource for innovations, in order to lead a company successfully, systematic handling of " knowledge", becomes more important. Today, the increase in value develops from the productivity and the innovation in business society.

What is the role of knowledge for individual? Knowledge is important for personal growth and development, knowledge sharpens our skills like reasoning and problem solving. A strong base of knowledge helps brains function more smoothly and effectively. We become smarter with the power of knowledge and solve problems more easily. Hence, innovation is about knowledge creating new possibilities through combining different knowledge sets. These can be in the form of knowledge about what is technically possible or not particular configuration of this world meet an articulated or latent need.

For Amazon knowledge innovation can bring the positive effects of technology improvement example, there are just a few of the ways in which technology may positively affect our physical and mental health. Health apps to track chronic ilnesses and communicate vitual information to doctors , health apps that anyone tracks diet, any kinds of sport exercise and mental health information . Hence, technololgy innovation had been applied to health apps smart phone product to help we to track our health from ourselves smart phone apps equipment easily any time.

What impact did this innovation have on daily life? It increased the regional differences among various groups of people across the country, it become a major method of long distance commnication to many years. It allowed people to sign documents from across the country. So, innovation can be applied to administive tasks aspect. Also, it can increase productivity and brings citizens new and better goods and services that improves our overall standard of living.

The benefits of inovation are sometimes slow to materialize. They often fell broadly across the entire population. According to Krathwohl (2002), he indicated that knowledge can be categorized into four types: (1) factual knowledge (2) conceptual knowledge (3) procedural knowledge and (4) metacognitive knowledge. So, if human hopes to implement knowledge innovation in success, we need to know how to learn these 4 types of knowledge innovation elements. For organizational knowledge management components example, the best four components are people, process, content/ IT and strategy.

Regardless of the industry size or knowledge needs of Amazon ecommerce organizations, organizations always need people to lead, sponsor, and support knowledge sharing. Sharing knowledge innovation how influences social development. Social innovation includes social processes of innovation, such as open source methods and techniques and also the innovation which have a social purpose, like activism , virtual volunteering, or distance learning. Social innovation may include the social processes of innovation. However, social innovation is important because it can provide a unique opportunity to step back from a narrow way of thinking about social enterprise, business engagement and to recognize instead the interconnectedness of various factors and stakeholders. For science and technolgical social innovation example, science and technology can help a nation's process and development science and technology innovation are connected with development because they have hostorical record of bringing advances that have led to healthier, longer, weathler and more productive lives and they are key ingredients to solutions to the most serious poverty and economic development challenges.

Social knolwedge innovation

So, social innovation may bring human benefits, such as providing food and lifestyle products world wide with focus on environmental and social innovation , giveing every child in the world the chance to learn code and helping the visually impaired interact with their surroundings. Hence, social innovation means new solutions (products, services, models, markets, processes etc.) that stimultaneously meet a social need (more effectively than existing solutions) and lead to new or improved capabilities and relationships and better use of assets and resources. However, a social innovation process consists of a sequence of activities that seals to find solutions to a special challenge. The process itself brings a new approach that has social impact in its means (process) and ends (solution).

There are 6 keys characteristics of a social innovation, as told by Colombian social enterprises, they include: Adresses real needs of people in a community, requires a deep understanding of the problem, localizes and humanizes the problem, builds trust and collaborates with the community in need, is sustainable and scalable and adapts constantly. Hence, every one may attempt to learn in social innovation. You will learn what social innovations are and understand how they are help solve societal problems. You will get an overview of important literature and debates on social innovation. You will also learn and apply methods to develop , implement and scale social innovations.

How can social innovation be improved to apply to Amazon ecommerce organization case? Major cross business/cross functional projects should also have social innovation objectives include: leadership programs that include volunteering activities may help employees develop their skills and lead to greater innovations during their daily work in any organizations. How does the idea of social innovation connect with social needs? We define social innovations as new approaches to addressing social needs. They are social in their means and in their ends. They engage and mobilize the beneficiaries and help the transform social relations by improving beneficiaries‘ access to power and resources.

For social innovation on education aspect, teaching technological literacy, critical thinking and problem, solving through science education gives students the skills and knowledge, they need to succeed in school and beyond. The essence of how science and technology contributes to society in the creation of new knowledge , and then utilization of that knowledge to boost the prosperity of human lives, and to solve the various issues facing society. Hence, when many consumers believe online shopping channel is one kind of convenient visa card payment and safe purchase , rapid goods transport to different countries, many different kinds of products choice channel. Consequently, global societies will be influenced to accept online shopping method / mode to replace traditional shops visting purchase mode. This kind of social online purchase knowledge innovation may influence Amazon ecommerce organization develops in success in long future days.

Amazon ecommerce organization warehouse products transport process and workplace innovation

How Amazon needs to consider warehouse goods delivery service tasks ?

Why does Amazon need to innovate its wareshoise products transport process to raise rapid products delivery service to different countries consumers as well as let buyers can view any products photos to feel their good quality, good design shape to attract customers consideration from its websites/ web stores in short time? What advantages will be bought to Amazon after Amazon warehouse delivery products had been innovated in success? Can products innovation help the product to improve its market image to Amazon? Can product innovation raise Amazon growth? I shall attempt to indicate reasons to explain above questions?

In our business societies, any products ought need to be concerned how to innovate them to be good quality. The key practical benefits of innovation may include: Improved productivity, reduced cots, increased competitivenesss, improveed brand recognition and value, building new business partners relationship, helping the business to increase turnover and improved profitability.

So, it seems that product innovation may bring positive impact more than negative impact to any businesses. In fact, after the buysiness innovates itself products, innovation ought may help the business to charge higher prices for new products before competitors products come on the market. Being innovative good for the firm's reputation, even people naturally interested in its future products, if they have been first in the past as well as innovations in processes can add value to existing products/ services.

So, the meant is by product innovation, it means that a product innovation is the introduction of a good or service that is new or significantly improved with respect to its characteristics of intended uses. Maninly, these reasons can explain why innovation is important: Innovation grows business, increasing profit, innovation helps any businesses stay ahead of the competitiion, innovation helps businesses take advantage of new technologies, such as Amazon ought need to innovate its warehouse products delivery service to let global buyers feel Amazon can provide rapid products delivery service to themselves homes after they paid visa card to buy any Amazon products from its web stores rapidly. On organizational benefit aspect to Amazon warehouse products delivery service innovation, innovation may help Amazon organization differentiates its 3 rd parties sellers themselves products when they decide to display their products photos on Amazon webstores, e.g. if your organization is using innovation on its processes, its because doing so will save your time, money, or other resources, and give your organization a competitive advantage over other companies stuck in their system.

In common, innovation may include four types. Incremental, disruptive , archihectural and radical, they help illustrate the various ways that companies can innovate. For technological innovative advantages, it increases productivity and brings citizens new and better goods and services that improve their goods and service that improve their overall standard of living.

The benefits of innovation are sometimes slow to materialize. They often broadly across the entire population. Hence, the advantage of product innovation may include: Growth, expansion and gaining a competitive advantage. A business that is capable of differentiating their product from other businesses in the same industry to large extent will be able to reap profit. Examples of product innovation in improved products involves introducing beter or more functionality to existing products, e.g. electric and gas lawn mower, GPs in car , battery car, non-manual driving auto car etc. So, product innovation is the creation, development and implementation, a new product, process and service, with the aim of improving efficiency, effectiveness or competitive advantages. Such as Amazon case, it needs to make products AI warehouse delivery service innovation, e.g. chaning to artificial intelligent products delivery service from workers delivery service in order to raise its goods can delivered to different countries buyers homes in the shortest time rapidly every day.

How does innovation help Amazon ecommerce organization grows its online buyers number? In fact, one of the major benefits of innovation is its contribution to economic growth. Simply put, innovation can lead to higher productivity, meaning that the same input generates a greater output. As productivity rises, more goods and services are producted. In other words, Amazon online buyers number grows.

Instead of innovation on product aspect, innovation is also important in the workplace, it can help staffs to raise efficiency. Innovation is vital in the workplace, because it gives companies an edge in penetrating markets faster and provides a better connection to developing markets, which can lead to bigger opportunities, especially in rich countries or developed countries.

However, when an organization decides to implement innovation before it implements , it needs to concern these possible risks of innovation. Operational risk, e.g. failing to meet your quality, cost or scheduling requirements, commercial risk, e.g. failing to attract enough customers, financial risk,, e.g. investing in unsuccessful innovation projects. But, when the organization decides to implement any innovation, it may hope innovation how contribute to success. Especially as customers become more demanding. Entrepreneurs need an innovation to survive to boost your business productivity, growth and profitability more easily.

IN simple, innovation may improve sales and customer relationship, reduce waste and costs boost your market position, improve employee relations. What does the right time to organizations make decisions to innovate ? When the firm discovers its customers use the product and they field praise and complaints. And they probably have ideas, however, that can be refined into a better product. Innovative companies make it clear they want ideas, that the door is open and there is always a friendly ear for changes, then they will start to come it. It may be right innovation time, when customers have any unique idea to concern the product after they use.

What is required to introduce innovation in an organization, such as Amazon ecommerce organization? To successfully implement innovation, you need to know exactly what makes an innovative organization as well as how it contributes to its growth. Our organizaitons also need to require an innovative culture where everyone is able to think independently . So, these reasons can explain why our organizations ought need to innovate, being bold in taking on the innovation challenge build people's readiness and receptivity to change, assisting people to resolve their unconscious biases and resistance to it, developing both customer intimacy and customer empathy. However, although innovation can bring positive impact in a business, but it is so difficult to implement because new ideas and initiative depend on the people who work for the organizations. It is a lot harder to achieve desired results. Innovation is not about optimizing gross margins , but about attempt how finding new ways to create more value for yourselves profit image to attract more customers consideration.

Innovation is difficult to implement, because no system, process, or industry knows how to change, more innovations are worth exploring for many. Technology for example, can be re-purposed into new innovative solutions provide your customers with new value. The challenge is that when it comes to disruptive innovation, it almost always involves " higher risk" compared to incremental changes and thus can not be managed the same way as regular business projects are managed.

Hence, maintaining quality and product improvement and process development often involves standardization, whereas innovation can rarely be standardized . Also, new innovationds can not be measured using the same metrics and value drivers as the existing products and services . To overcome this major barrier to innovation, companies should approach disruptive innovation differently compared to how they are used to approaching regular projects. This, any organizations need to understand that they may fail to innovate. In the beginning, innovation and most specifically disruptive kind, is inferior to the existing products and services on the market. Because product improvement takes a lot of time and requires multiple iterations, the value for the customer at this point is minimal , when distuptive innovation initally caters only to a small and not so profitable customer base, established organizations are focused on serving more demanding, high and customers using their existing value channels. This is where it typically has higher profit margins, which is why established companies with rational decision-making processes usually choose not to invest in disruptive initiatives in the easrly stages. The problem ocurs when incumbents attempt to apply new technologies to trheir existing value networks or refuse moving into new markets because they are seen as too small to drive growth goals or are simply perceived to have too low margins. So, for organizations that prioritize reaching scale through operational efficiency, it makes more sense to focus on growing the business through incremental means, such as invest in risky and uncertain innovations.

On conclusion, in reality, however, Amazon ecommerce organization needs to do both simultaneously improve the core business and exploit new business opportunities . Hence, when Amazon organization decides to innovate its 3 rd party sellers their product, on warehouse goods AI delivery processing or AI workplace safe working environment . Amazon needs to find a balance between different types of innovations to choose which is the most effective goods delivery and warehouse workplace safe and efficient goods delivery innovation, which is a lot more sustainable may be stay in the Amazon e-business growth in the long term to in order to implement Amazon AI warehouse goods delivery service and warehouse working environment efficient and safe workplace innovation in order to let its all warehouse workers to feel comfortable and safe feeling and raise their goods delivery tasks efficiently.

AI creates Amazon efficient service growth

How does (AI) technological development will influence Amazon develops in success ?

(AI) robotic automation technology can bring Amazon warehouse goods delivery service task more efficient

Future, (AI) robotic automation technology can be applied to Amazon warehouse goods delivery tasks in order to help it to shorten goods transport time to be delivered to overseas different countries buyers homes in the shortest time rapidly every day. For example, nowadays, UK computer and space explore technology had reached the mature stage. It means that UK government ought not need to continue spend much resource to research these two kind technologies. Otherwise, the (AI)robotic automatic manufacturing technology, e.g. human intelligence new product. It has need to develop because human intelligence machines will bring beneficial to satisfy human everyday life need, e.g. hospital patients' activities need, if the patent who can not walk easily, but the human intelligence machine can assist the patient walk to anywhere conveniently. So, he/she does not need to sit on wheel chair and apply the human intelligence machine man to help him/her to drive on the intelligence automatic driving vehicle to go to anywhere conveniently.

Otherwise, increased automation in low wage countries, e.g. China, Korea, Africa, Hong Kong etc. which have traditionally manufacturing firms, could use automatic technological manufacturing to bring lose cost advantage and potentially lose their ability of achieving rapid economy growth by shifting workers to factory jobs. So, UK government and businessmen needs to consider automation technology development, i.e. 3D printing manufacturing industry will encourage UK companies to move manufacturing process, closer to gain the biggest advantage from this 3D automation technology development.

A growing concern of premature de-industrialization in energy and developing countries could require new models and a need un-skillful the UK workforce. In the future, the best way toward for UK cities will reduce their exposure to automation is to boost their technological dynamic and attract more UK skilled workers. Automation technology progress can give UK manufacturers' employee benefits, such as long term healthy productivity improvement, raising productivity efficiency and product quality, macroeconomic and microeconomic effects of automation technological change, it's change will be beneficial to UK society, i.e. automation active labor market policies, which could help UK job seekers find jobs from training to incentive to support self-employment to create high technological job employment chance in UK society. So, raising science, technology, engineering and math subjects update skills level are needed to UK any universities, which can be increasingly important in UK society, these factors could complicate the ability of UK high automation technology education to adopt to the UK automation manufacturing technological change. A talent mismatch already exists in UK, with many well UK educated workers can find employment in lower-skilled jobs. To combat this, greater coordination will be needed between the education, training and employment sectors in UK society.

Why are high automatic technology product development models needed to improve to Amazon warehouse goods delivery improvement service ?

UK government and manufacturers need to consider how to achieve high technology product development models. According to Hauser et al. (2006) indicated the high technology (high tech.) development process, is influenced by the innovative process, bringing products on exception value which stimulate product market demand. Innovation provides products the specific basis for which world economies compete with each other on the global market. Able to find new solutions, innovations generate significant changes in existing markets, destroy them, or create new marketing (Hauser et al. 2006). So, UK manufacturers need to concern on any manufacturing high technology product development process because which can influence any new products development to manufacture to sell to any overseas or domestic both markets successfully.

What does high tech. product warehouse goods delivery transport service improvement mean? Mohr et al. (2010) argues that there are two reasons why it is important to clarify and specific high technology : (1) due to the impact of technologies on the economy, attempts are made to classify economic production and incomes ; (2) due to the impact of high tech. on the environment. Standard marketing strategies are being modified and adopted , therefore, it is necessary to know the products to focus on. Why does Amazon need to consider warehouse high technological product delivery transportation process? Nowadays, high tech. products are complex, advanced, requiring specific technical knowledge, which is technologically not discontinued and being produced at the companies which have twice as many technical personnel and invest twice as many in scientific research and development than other companies. Moreover, these products are time-sensitive as scientists are continuously searching for new approaches for invention of more advanced technologies which make all preceding ones lower-ranking. The most important, nowadays global consumers will adopt the particular technology. It means that Amazon needs to improve high technological goods delivery service process to avoid the delay when global any online shoppers choose to buy any kinds of high technological products from Amazon web stores.

Anyway, nowadays customer individual needs in high tech. environments are characterized by sudden changes related to unpredictable fashion. Even, consumers concern about how to preserve new product' competitive technological standard is completely incompatible with technological uncertainty. The most important factor is the prevalence rate of any new products development process, which is influenced by slower than of traditional products. In many cases high-tech. automatic product market are being materialized slower than which are expected. The technological uncertainty challenges will exist in development process, such as uncertainty related to the timetable for development of the question whether the new product will be function as promised. In automatic high-tech. industries, the time requires for product development is difficult to predict as , commonly, it takes longer than expected , uncertainty related to unanticipated consequences and uncertainty about the product life cycle related to competition products. In conclusion, these factors will influence future Amazon any kinds of new automatic technological product warehouse delivery process to avoid to delay to deliver to different countries buyers themselves homes rapidly every day.

Future economists predict automatic technology how to influence future ecommerce goods efficient delivery service

Before, all over the world presented picture of demonstrate in London on the occasion of the meeting of the G20. Some economists indicated disastrous economy consequences will occur to any one of Western country , such as UK, so if any one of Western country did not consider automatic technology development to itself country. They indicated one example, such as material incentives to produce disappeared throughout Russia and, when Society leadership called off the experiment, the country faced industrial output reduced to 10% of what had been registered in 1914 and agricultural output reduced to such low levels as to cause widespread famine.

Why would UK encounter disastrous economy consequences if UK government did not encourage manufacturers spend money to invest to innovate automatic technology industry, such as Amazon online goods global delivery service case? According to a variety of anthropological studies, a collectivity is unable to operate efficiently with everybody giving talent workers have chance to devote whose best effort to manufacture any high technological products, e.g. human intelligence vehicle or airplane. Hence, economic incentives are needed to manufacturers to invest high technological automatic industry development. Because the economists predict future ecommerce organizations will have many talent worker numbers, their number will be more than a certain number of normal effort workers, due to technological education level is very excellent to provide to train many young technological manufacturing students to find this kind of high technological manufacturing job. So, the high technological manufacturing job seekers need to future ecommerce market in order to raise ecommerce any organizational development.

Assuming that future UK ecommerce market will have high technological automatic manufacturing workers who would desire only to introduce changes in the workings of the international e-commerce economic order and policies of countries participating in the present economic order rather than change the order itself, what will be UK manufacturers their specific economic preferences in the future? It implies tnat either concentrate on spending more investment to automatic high technological development, e.g. human intelligence automatic high technological products or still concentrate on spending more investment to common traditional technological products to future UK ecommerce market.

However, UK was a developed Western country which had had strong automatic high technological development effort very long time. Otherwise, it compared to some developing countries, such as Asian China, Hong Kong, Korea etc. Asian countries their future economic growth rate will show un- surprising , different patterns, so the Asian countries has weak effort to invest high automatic technological product development, such as human intelligence technological development. The catching-up process suggests low economic growth rate in the high automatic technological product development to the Asian developing countries in the future.

Hence, the future economists predict that it views as probable successors of the Western world economic leadership if any Western country , such as UK manufacturers who prefer to invest to any high automatic technological products development , e.g. developing on human intelligence automatic technological products more than traditional common technological products development. On the one side, but it seems important to stress that two very poor countries among the challengers-China and India-are examples of countries that changed their institutions and economic policies from no or little economic freedom to more economic freedom. Because there two countries whose governments prefer to lend loans to encourage their country manufacturers prefer to invest high automatic technological products manufacturing. On the other side, attitudes toward foreign direct investment (FDI) have undergone change since the 1960 s and a large majority of less developed countries, e.g. China and India are now competing strongly among themselves and with developed market economies for direct investment from multinational companies. So, UK will face China and India high automatic technological product competitors in the future. And in fact, all countries that joined Western developed economies did that without much (if any) external inflow of public resources. It is right time that UK government needs to lend loans to encourage domestic manufacturers to invest high automatic technological products to raise whose international high technological products sale effort to win its future competitors. So, machine resources will be increased demand to o UK manufacturers if who chose to spend machine resources to innovate to manufacture any new and high technological automatic products to raise human daily life needs in the future. It means that it is right time UK manufacturers need buy much machines to prepare to manufacture many future high technological automatic products when these machine prices are low. Because the future global machine prices will possible be raised if many China and India manufacturers will also buy many machines in the future. For example, USA government had provided much financial support to assist sugar cane producers to develop their businesses. And they are dependent to a much larger extent than sugar cane producers and sugar processors in the USA on government. Without very high subsidies to renewable energy generation, they would not have survived at all. So, USA government had been the first country which could lent much financial assistance to encourage domestic renewable energy generation manufacturers to develop high technological energy manufacturing business. So, UK government needs follow USA to lend financial assistance to encourage domestic high technological automatic industry development. Future economists also predict China and India will be competitors for future leadership in the global e-commerce economy, special high technological products. China has been the media and analyst's favorite for quite some time. Quantitative projections have seemingly supported such expectation. Such as China and India had manufactured many high technological new space rockets products, ocean war large ships etc. Moreover, China has become one of the major world trade players in the early twenty-first century.

Many long-term forecasts, assuming similarly high economic growth rates in the decades ahead, predict that China will surpass the USA in terms of aggregate GDP somewhere between 2020 and 2030 or later, say between 2030 and 2050 year. The future economists conclude on the basis of these predictions that China will not only pass the USA in aggregate product (GDP), but its economy and economic policies will influence the rest of the world to a similar extent that the USA does at present.

I stressed a very important point to future global ecommerce organization growth, namely that the UK future high technological automatic product competitor China and India, namely that economies not only grow, but in the process change their structure. China and India have been industry very rapidly (the first transition) and building the physical infrastructure that accompanies industrialization changes to technology in the future. However, at a certain per capita GNP level the two countries, such as China and India will face another structural shift when which technological development will reach the mature stage in the future. China and India had been primarily historical pattern of economic development because the shift in the role of engine of growth from industry to services is to a much greater extent a qualitative shift. Both higher and different skills are required. And, even more importantly, interactions generating ideas driving the highly human-capital-intensive service economy require a much freer environment, not only in the economic area. Chinese exports have been heavily labor-intensive. This being the case, they contributed to the expansion of industrial employment, offering for the first time in the history of China a taste of (very modest) prosperity to more than 100 million new industrial workers and their families. This is the major component of the success accomplished by Chinese economic growth. Richer trade partners create room for more trade, so the Chinese should hope that intra-South trade, that is, trade between the emerging economies of Asia, the Middle East, Africa and Latin America, will open up new and growing opportunities. I presume that if Western economy , such as UK did not developed high technological automatic industry to stable their social welfare, so thoroughly slowed down their economic growth.

Will it allow China to accomplish the transition to a mature, innovation, service-sector-based ecommerce market economy? It has allowed the economy to industrialize much more successfully, even if the labor shift from agriculture to industry has not yet been completed. But it is a long way off the next major test: the second high technological industry transition of the economic structure to China. Bear in mind that Russia attempted it twice and failed at both attempts.

But even, assuming that China at some point in the future does succeed in accomplishing the second transition, will it be able to supersede the USA, for example, as the main global high automatic technological innovation center if it wants to become the No.1 global high technological industry ecommerce economy? Given the nature of the centralized state and its stability to collect financial resources , China's ability to increase research and development expenditure to high automatic technological products and to hire a mass of researchers, engineers, technicians and other specialists should not be doubted. This process in already taking place.

But , again, Soviet Russia already exceed the USA in the R&D/GDP ratio in the 1970s, long before the communist collapse, with no effects on its innovativeness. Inputs matter less than outputs, quantity in the innovation process mean much less than quality. The latter characteristics depends importantly on economic, civic and even political institutions. Otherwise, independent India had three options open to it in 1946s. It could pursue spontaneous economic development, with some state intervention to be sure, along the lines of basically free market capitalism; it could turn the clock back and try to recreate the rural-agricultural and handicraft based. The dominant way of thinking was Society -style priority to industrialization and , within industralization , priority to heavy industry. In other words, not textiles and clothing, which has been developing well in India since the mid- nine teen century, but production of sewing machines and , even better, production of machines the produce sewing machines.

The results were only to be expected. The heavy stress on the expansion of capital-intensive heavy industries in a very poor country quickly strained the ability of the Indian economy to generate adequate savings. Moreover, some of these industries were above the level of industrial competence of an underdeveloped economy. Thus, the amount of required resources (capital, skilled labor) was usually larger per unit of output than in the same industries in more mature, richer industries economies. In another view point, India will develop light industries, just as any other poor country with a great deal of unskilled labor, had a comparative advantage and no less importantly, an economy in which, due to their low capital/labor ratio, light industries could employ many more people, spreading prosperity more widely in a poor country. So, it explain that why China will have more effort to develop heavy high technological industry in the future. Thus, India got less economic efficiency, less employment than in a spontaneously developing economy, less ability to compete internationally in light industries suitable for an underdeveloped economy and finally got heavy industry unable to compete even on the domestic market and, therefore requiring no less heavy a dose of protection. Overall India got an underperforming economy, in particular in its relations with the rest of the world.

To conclude by comparing the performance of the traditional sectors of the Indian economy and the performance of its modern, human -capital-intensive subsector of manufacturing and skill intensive service sector. The latter both employ workers with high-and medium -high skillful level (in branches ranging from computer software and biotechnology and pharmaceutical high technological light industry). India is ahead of China in terms of the output and export of such products and services. Thus, it implies that Amazon ecommerce organization ought concentrate on developing high automatic heavy high technological warehouse goods delivery service industry, e.g. human intelligence technological warehouse delivery products because these industry is not better development to other many countries' strong effort , such China and India large population countries, they still have weakness to develop warehouse AI delivery service skills in themselves ecommerce organizations.

Amazon succesful leadership management behavior

How Amazon leader be one successful CEO?

If it is not, what are the unique personal characteristics between one general CEO and one successful CEO ? Instead of personal characteristics, which kinds of skills to any successful CEOs, they will need? Why do some CEOs encounter fail to manage themselves companies? Why do some CEOs feel difficulties to manage themselves companies? This book can indicate some evidences to explain what factors can influence the person can be successful CEO to let readers to understand. Readers can learn whether what personal factors can help general CEO to become excellent CEO in any organizations.

The different characteristics between common CEO and successful CEO

The characteristics of common CEO

What are the common characteristics to general CEO? IN general, CEO person specification includes: High level of self-motivation, creativeness high level of self confidence. So, In general, qualitiies and traits of a chief executive officer, he /she may have courage, passion , but an excellent CEO is draw to change and effective action. Also CEO needs have resillence and drive ability. It means theat the CEOs, leaders ought know that taking risks and making large-scale changes can lead to organizational growth or can fall dramatically.

A CEO must posses certain traits to be an effective leader, e.g. ability to learn from the past experience, strong communication skills, buildinh relationship, realistic optimism, easily understanding, listening people and adapting to necessary managment styles.

What makes a good CEO leader? At a leader, a common CEO needs to posses strong communication skills. From motivating your employees to meeting set deadlines. You should have the ability to communicate your needs, when you hope to become a common CEO. If you want to be a good CEO, you must be consistently clear in your communication. But, some succesful executives may have these 7 perdsonality traits, such as visioning, in-depth problem solving and analysis, attapting change, driving for results, influencing and persuading, managing others, organizational resources allocation.

Hence, it seems that it is not all people these 7 successful Ceo. In fact, many companies own common CEOs more than successful CEOs in human resource view. In general, an organizational leader (CEO) , he /she does not need special management train, he /she depends on his/her past working experiences training to climb up to become the company's leader. SO, taking risks and acceptance fail or acceptance attempt, they may be general CEO personal characteristics . All of these personal characteristics, it is not difficult to own to general CEOs in any organizations. But if the organization hopes to help the business owner to manage himslef/herself overall organizational different departmental operation more effectively and efficiently. The CEO must own unique personal characteristics and excellent managing ability in order to gelp his/her boss to manage whole organization excellently. Hence, one successful CEO must own unique personal skills or abilities and personal judgement characteristics to compare common or general CEOs in any organizations.

For example, when one organization has serios financial resource allocation to provide different departments challenges, it is the best chance to examine the CEO how to apply shortage financial resource to provide to different departments in order to still keep efficient operation aim. If the CEO can know how to arrange shortage financial resource to provide to salespeople salaries expenditure , shop or office rent, water , electricitiy , telephone fee etc operational expenditure, factory workers salaries and factory machines productive energy expenditure, product manufacturing processing material expenditure etc. resource managment expenditure allocation effectively. Consequently, the overall organizational performance can still keep positive growth, or profit can still raise. Then, I believe that this organization leader may be one successful CEO , he/she does not ne one common CEO or failure CEO role to this organization.

The characteristics of successful CEO

What factors cause the CEO can be one successful leader to his/her organization? What are this CEO personal unique characteristics own? I believe that any successful CEOs must oen unique personal characteristics that common CEOs must not own, I shall explain as below:

Usually , any successful CEO must need time from common CEO to become. Every successful CEO must communicate with their employees using concise, easy-to-understand language, open-mindedness, approachability, growth mindset, ethics, decisiveness. All these characteristics may need to any successful CEO. A chief executive officer (CEO) is the highest-ranking executive in a company , whose primary responsibilities include: making major corporate decisions, managing the overall operations and resources of a company, acting as the main point of communication between the board of directors (the board) and these corporate .

Hence, a successful CEO needs have these 5 key managerial skills: Technical skills, conceptional skills, interpersonal and communication skills, decision-making skills. The roles that a manager plays in the organization require having some skills. Hence, one successful CEO must need to know how to supervise and manage stsffs for his / her overall organization. A successful CEO also needs to understand every part and function of the business: accounting, finance, HR, marketing , legal , operation supply chain, sales and information technology. Also, one successful CEO also needs to consider organizational cultural fit, industry understanding , building good soft communication skills between staffs and him/her or between customers and its sale service staffs. Moreover, a successful leader with a CEO mindset has a clear direction for the future, and is not afraid to share it. Do not be scared to set yourself, when you are the organizational CEO and your team, ambitious and exciting goals, sure that you have smaller, achieveable steps in these as well. So you can maintain motivation.

For next example, if you are your firm's CEO, you can help your firm shareholders grew in power and their demand for booming stock prices led to booming pay. It means that you 's CEO salaries increase or decrease, it depends on share price driven salaries, when your organization's share price can often keep high price position to compare similar competitors ' share prices. Then, you 's CEO salary may keep increase, because your can manage firm's share price often keeps on high price position in share market. However, one successful CEO must need to own there managerial

skills in order to help his/her organization can grow rapidly. The managerial skills may include as below:
Technical skill, it means that the abilities, knowledge, or expertise required to perform specific, job-related tasks. Technical skills are related to jobs in science, engineering, technological, manufacturing or finance. They are learnt through on-the-job experience or structured learning, e.g. data analysis, project management, technical writing, software proficiency, programming languages, artificial intelligence, machine learning, data engineering, visualization, network and information security, cloud computing.
The next is conceptual skill, a successful CEO also needs have conceptual skill, it is the ability to analyze and evaluate whether a company is achieving its goals and its business plan. Conceptual skills are skills that enable individuals to identify , conceptualize and solve problems. It is important in the workplace because it allows professionals to think and woth though abstract ideas and come up with multiple solutions to complex issues. Hence, conceptual skills include the ability to view the organization as a whole, understand how the various parts are interdepentend, and assess how the organization relates to its external environment. These skills allow managers to evaluate situations and develop alternative courses of actions. Conceptual skills may include: abstract thinking, analytical skills, congnitive skills, communication, contextualizinf, creative thinking, critical thinking, decision making. Hence, one successful CEO may be a conceptual person, who is one conceptual thinker, he /she has an understanding of why something is being done. The conceptual thinker can think at an abstract level and easily applythe CEO himself/herself insights to the suitation. So, any comon CEO may attempt to improve conceptual thought processing and increase work performance by these methods:
Observe leadership. using challenges as case studies, seeking outside knowledge, staying up-to-date on the industry, applying new practices, disucssing concepts with colleagues, finding a mentor, learning about the organization. So, one successful CEO needs to own critical thingking skills: Analysis, interpretation, inference, explanation, self-regulation, open-mindedness and problem solving, strategic thinking includes careful and deliberate anticipation of threats to guard against and opportunities to pursue.
Ultimately strategic thinking and analysis can help CEO to lead to a cleear set of goals, plans, and new ideas have abstract thinking and feeling, it is the

ability to understand concepts that are real, such as freedom. So, analytical skills refer to the ability to collect and analyze information , problem solve and make decisions. Successful CEO can know how to use analytical skils when detecting patterns, brainstorming, observing , interpreting data and making decisions based on the multiple tailors and options available to you, such as the organizational CEO, e.g. creative thinking visual art, communication skills and open-mindedness to any one successful CEO, he /she ought own.

Finally, successful CEO needs have excellent interpersonal communication and decision making skill. Why is decision making and communication an important skill? It can help any CEo to raise the ability to make a decision of good leadership skills. Decision making is an on-going process in every organization, large or small. Having critical thinking skills allow the CEO to ascertain the problem and come up with a solution that is benefical to the company and its employees. So, one successful CEO needs to own soft and hard skils both.

For making decision for the organization, the successful CEO must have decision making skill to find the best solution to the challenge in process. He /she can define the problem, challenge or opportunity clearly, generate of possible solutions or responses, evaluates the costs and benefits or pros and associated with each option, selects a solution or response and knows how to implement the option chosen clearly.

Any CEO can learn how to improve decision making in workplace, such as following these steps: starting with the desired outcome, or goals, rely on data and insights to spot patterns, use S.W.O.T analysis, simulate the outcomes, trust your instincts and identify your cognitive biases. Identifying critical factors which will affect the outcome of a decision, evaluate options accurately and establish priorities, anticipate outcomes and see logical consequences, navigate risk and uncertainty,, reason well in requiring quantitative analysis. On conclusion any one CEO must need time to improve his/her hard and soft skills in order to become one successful CEO to his /her organization.

What skills to one successful CEO owns

We can learn that it is difference between one common CEO and unique successful CEO personal characteristics. Then, it beings this question: What are the skills that one successful CEO ought own? I believe that one successful CEO ought own these skills that one common CEO he /she won't own.

Building excellent communication skill:

Top performing CEO, ough know that strong communication skills are the secret to influence final success. Successful CEOs understand that influence is required if they are to inspire people to willingly act upon what they have to say. Highly influential CEOs deliver on these communication skills daily as below:

Successful CEOs understand the importance of clear, concise communication. They recognize that in the absence of simplicity comes confusion. For example, one organizational CEO sent an email to his employees listing only three objectives he wished for the company to focus on : " customers, team and execution". Instead of providing a dozen areas of opportunity . This CEO also maintained a short list of objectives that were clear. He kept employee goals concise, guiding them to focus to perform excellent services to let customers feel satisfactory. Such as this case, it explains that where one CEO focuses on concise and clear communication, he makes it easier for others to follow. This level of influence ensures that others remember what was said and are inspired to act accordingly.

Top CEOs are known for their sharp minds, and business acumen. They know that frequent communication between employees and senior leadership to be very important in their ability to stay engaged. For conference discussion case, when a successful CEo discusses ongoing company objectives and action items. he begins each all by sharing status updates of previous discussions, including what the executive staff is doing to secure the company's future. Each call concludes with a 30 minute open question forum, where all empllyees have the chance to ask questions and on ideas the executives discussed. Allowing employees to collaborate and share ideas creates a sense of ownership . It permits insight into the company's goals and encourage employees to engage in its success with ideas of their own.

By creating an atmosphere, successful CEOs encourage employees to share ideas. As a result, CEOs build stronger relationships and deepen trust in leadership . For example, when one CEO maintains an open-door policy and is known for frequently visiting employees on the floor, no matter the department or position. As a result, the CEO builds personal relationships that foster trust and candidacy that only comes with real influence. So , the difference between the successful CEO's communication skill and common CEO's communication skill is that a common CEO has influence based solely on his /her title may intimidate employees to act on direction, but successful CEO is a leader who influences others to act willingly has establish the trust and credibility necessary for lasting success. Unfortunately, too may ledaers fail to share this level of detail with their staffs leading employees to question their intention. They neglect communication weaknesses that need improvement . As a result, the entire organization will benefit from improved performance and communication . When leaders admit they are not perfect and are willing to improve, employees follow suit.

What makes a CEO successful? Findings from a database of 17,000 c-suite assessments reveal that successful CEOs demonstrate four specific behaviors that prove critical to their performance. They are decisive, they engage for impact, they adapt proactively, and they delive reliably. So, as a leader, you need to posses strong communication skills. From motivating your employees to meeting set deadlines, you should have the ability to communicate your needs and even show your employees how things are down. If you want to be a good CEO, you must be consistently clear in your communication.

What is the most important skill to a CEO? In short, the single important role of a CEO is to make absolutely certain that the right CEO is running the company and then do what is necessary to encourage that CEO's effectiveness, strategy, vision, culture shareholder value, all crucial and all within the scopr of the CEO's role. Hence, successful CEO needs to know whether what he/ she actually does. A chief executive officer (CEO) is the highest ranking executive in a company, whose primary responsibilities include making major corporate decisions, managing the overall operations and resources of a company, acting as the main point of communication between the board of directors. Hence, CEO needs to report to the board of directors, with most CEOs being members and sometimes chair of the board, president, they report to the CEO and the board of directors and cometimes they are board members.

Hence, a CEO needs to understand every part and function of the business: accounting, finance, HR, marketing, legal , operations, supply chain , sales , information technology . In business speak, the CEO's job is to define the mission (purpose), strategy (direction), and metric (pace and performance). These three elements provide the essential elements that a growing company needs to be able to perform. So, a successful CEO must need to find the effective strategy to help his/her company to solve any challenges in ay time. When the chairman technically has higher level power, the CEO is indeed the boss of a company. The CEO does by the law answer to their board of directors, which is ultimately headed by the chairman. In general., the CEO job starts when the organization reaches about 20 employees, prior to 20 employees, the job resembles more of a product management role. CEOs at this stage are trying to develop a visable product and generate some revenue.

In general, successful chief executives tend to demonstrate four specific behaviors that prove critical to their performance. For example, holding people accountable and the ability to motivate a team, high-performing (CEO) do not necessarily stand our for making great decisions all the time, rather they stand our for being more decisive. They make decisions earlier, faster, and with greater conviction. Also, they do so consistently, even with incomplete information, and in unfamiliar domains . Interestingly, the highest IQ executives , they are intellectual complexity, when the quality of their decisions is often good, because of their pursuit of the perfect answer, they can take tool ong to make choices or set clear priorities and their teams pay a high price. These smart but slow decision makers, their teams either grow frustrated , which can lead to the attrition of valuable talent ot become overcautious themselves.

Moreover, high-performing CEOs understand that a wrong decision is often better than no decision at all. It means that a bad decision was better than a lack of direction. Most decisions can be undone, but a successful CEO has learnt to move with the right amount of speed. To that end, successful CEOs also knows when not to decide, whether a decision should actually be more lower down in the organization and if delaying, it is a week or a month time, would allow important information without causing harm.

Hence, strong performers balance keen insight into their stakeholders‘ priorities with focus on delivering business results. They start by developing an understand of their stakeholders' needs and motivations and get many people on board by driving for performance. So, CEO needs to bring others along plan and execute disciplined communications and influencing strategies. Indeed, the skilled CEO gains the support of their colleagues by confidence that they will lead the team to sucess, even if that means taking uncomfortable or unpopular moves. These CEOs do not shy away from conflict in the pursuit of business goals. The ability to handle different viewpoints significantly faster than average.

Factors Influence CEO Success Or Fail

In fact, any CEOs will be influenced to succeed or how how to manage their organizations by personal psychology and personal skills or knowledge and external environment factors. However, CEO is such leader role to any organizations. Can owning high level leadership skillful CEOs manage their organizations more eaily to compare owning low level leadership skillful CEOs? May leadership skill be main factor to influence any one CEO's person success or fail? I shall attempt to explain as below?

In fact, I feel any CEo must need have these kinds of leadership skills. They ma y include: First, recognizing strengths, everybody with an organization has their strengths, and it is essential that you are able to identify the strengths of individuals, and recognizing weaknesses, second, reacting to employee needs, third, clarity and fourth, willingness to make tough decisions and conflict managment skill. Why does every CEO need have leadership skill? Effective leaders have the ability to communicate well, motivate their team, handle and delegate responsibilities, listen to feedback and have the flexibility to solve problems in an ever changing workplace. Hence , in CEO personal psychological view, an excellent CEO is drawn to change and effective action, courage , passion and resilence and drive attitude . A good leader knows that taking risks and making large-sacle changes can lead to organizational growth or can fall dramatically.

Why is CEO leadership important to the performance of a firm? Research evidence provides overall support for the positive relationship between leadership and firm performance (Lowe et al, 1996). CEOs with transactional leadership can successfully manage goal accomplishment and contribute to the enhancement of the firm performance. So, it seems that CEO's leadership has indirect or direct relationship to inflience his / her whole organizational performance whether it can grow up rapidly or slowly, even fall fown rapidly or slowly, because if the firm's CEO can not manage different department teams cooperate efficiently. Then, the firm's leader personal leadership effort may influence whole organization different department teams to cooperate smoothly. Although, CEO must not need to contact all department staffs every day, but he /she needs to contact any department managers in order to know whether their departments have any challenges , when managers feel difficulties to solve. If the CEO had low level leadership ability, he /she won't discuss with any department managers to conclude the best opinions to solve the challenges more easily. For example, when financial budget department manager discovered that his organization had deficit challenge recently. If he can not solve lack of cash available problem. Then, his organization's deficit will be increased easily. Consequently, it will influence staffs salaries can not pay on time, it won' t have cash to buy enough manufacturing material to prepare to be supplied to manufacture to provide products to provide to market to sell, factory can not have enough money to buy new productive machines to replace old productive machines etc. different kinds of organizational operational needs.

However, if the CEO owning high level leadership skills, he may know hoe to lead this financial budget manager to solve " deficit" challenge. Although, deficit seems to be simple matter, but if this financial budget manager can not know how to manage cash available in order to allocate to different departments operations, e.g. allocating the limiting amount of cash to urgent departments to use in prior. Then, deficit challenge will increase cash shortage number more seriously. So, high leadership skill mist need to any one organizational CEO in order to help his/her organization can solve any challenges more efficiently and easily. Otherwise, low leadership effort owning CEO only influences his/her organization can not grow up more rapidly, even it can fall down rapidly. Consequently, the organization will only liquidate or it will be sold out rapidly.

What does excellent leadership skill need to CEO? Many psychologists indicate excellent leadership may include these essential elements: Integrity, ability to delegate, communication, self - awareness, gratitude, learning agility, influence effort, empathy.In common, characteristics of a good leader, he / she can help staffs and makes the essential large-sacle decisions that keep the organization can operate efficiently and reduce challenges occur. Integrity is especially important for top-level executives who are charting the organization's actions and making countless other significant decisions. Ability to delegate , delegating is one of the core responsibilities of a leader, the goal enables the CEO's direct reports, facilitate teamwork, provide autonomy, lead to better decision making and help the CEO's direct reports grow. Effective leadership and effective communication are intertwined.

So, any CEOs need to be able to communicate in a variety of ways, from transmitting information to coaching your staffs (managers). Any CEO must be able to listen to , and communicate with, a wide range of staffs across roles, social identities and more. The quality and effectiveness of communication across the CEO's organization directly affect trhe success of the CEO's business strategy. So, better communication skill can actually improve the CEO's organizational culture. Self-awareness is focuses skills for leadership. The better leadership skill to the CEO , he / she can understand himself / herself managing ability, the more effective , he /she can do. Do you know how other people view you or how you show up at work?

Gratitude can lead to higher self-esteem, reduced depression and anxiety, learning agility is the ability to know what to do when you do not know what to do. So, great leaders are great learners, with strong learning agility to get started , when you are one organization's CEO in beginning, you must need to learn how to influence your whole organizational staffs behaviors to be the perfect.

" Influence" may be through locial , emotional or cooperative appeals, is a component of being an effective leader, influence is quite different from manipulation, and it needs to be done and it requires emotional intelligence and trust. Empathy is correlated with job performance and is a critical part of emotional intelligence and leadership effectiveness. A successful leadership CEO needs have empathetic behaviors towards his / her direct reports, empathy can be for improving workplace conditions. Courage is such that when the CEO wants to voice a new idea, provides feedback to a direct report or flag a concern for someone above him / her. That is part of the reason courage is a key skill for good leaders. Rather than avoiding problems or allowing conflicts to faster, courage enables leaders to step up and move things in the direct direction. A workplace with high levels of psychological safety and a strong coaching culture will further support truth and courage. Finally CEO needs to know how to treat people with respect on a daily basis is one of the most important things . A leader can do that it will ease tensions and conflict, create trust, and improve effectiveness.

On conclusion, when the CEO can know how to build these psychological and emotion feeling, then the CEO can be trained to improve leadership skill in order to know hoew to manage his / her organization efficiently and effectively.

Can training provision raise CEO leadership

Any organizational CEO is the top level managerial position. So, it is one job to anyone. Any organizational positions may have training provision in order to improve the staff personal job skills, e.g. organizations can provide internal accounting training to accounting clerk, even accounting manager training in order to let they can learn company's accounting policy in order to improve their accounting tasks more proficient, or a law firm can provide legal draft training to general law clerks in order to improve their legal draft writing skils, or one company's data processing department can provide data processing training to improve data processor typing speed to learn more proficient to type its documents, or property agent firm can provide property sale speaking training to its property sale agents to improve their property sales presentation skill or insurance agent firms can provide insurance sale speaking training to its insurance sale agents to improve their insurance sales skills.

Hence, it seems that any organizational positions may apply training methods to improve any staff individual performance in any organizations. So, it brings this question: Can organizations provide training to improve CEO performance to achieve more proficient? To answer this question, we need to suppose same kinds of business CEO positions , they ought be trained to improve their performance. If it is true, whether what kinds of businesses CEO positions, they may be applied training method to improve their managerial skills? How and why to these kinds of business CEO positions, they can be trained to improve their proficiency? I shall attempt to answer as below:

In fact, general CEO typically have a bachelor's or master's degree in business administration or a fiels related to their industry. Some CEO positions require that candidates have a master's or even depending on the industry , education, for instance. So, one university CEO or president, he /she needs have education psychology master or doctorate level to any kinds of degree. They need have high educational level to do university leader. Hence, CEO, training can focus on learning or educational aspect, e.g. certified CEO program is general education certification course designed for business leaders globally (CEO senior managers) and aspiring business leaders who are looking to build upon existing qualifications and business experience.

What does CEO coach training method mean?

However, every CEO needs a coach in organization , because a coach suports the CEO to manage conflict effectively often decisoin of the CEO please one group and displease another. The CEO needs a partner who the CEO can be open with, one who is going to be sensitive, and objective, honest and respectful. Hence in any organizations, a coash can teach the CEO these managing skills to manage their organizations more easily, e.g. flexibility, value driven decision making, delegating, leadership, clear vision implementation.

Hence, in general, coach training needs ususally take 2 to 3 years to complete. The organizational CEO will does on the-job-training and spend time with a training provider. Employers will set their own entry requirements. So, in popular, many CEOs use a CEO coach over the course of their career. No athlete would be embarrassed they use a coach. Yet, CEOs believe they do not need a coach of their own. However, any organization is the final deicion making person, when the organization feels that the CEO's managerial effort is poor, it can attempt to provide a proficient managing experience coach to train this CEO's management skill in order to achieve this organization's CEO ability requirement within 2 to 3 years. For example, a CEO peer group training, interchangably called CEO peer groups or networks, these organizations generally arrange reqular meetings in confidential environments where CEOs can share ideas, best practices, experiences and advice together to attempt how to improve their managerial skills.

So, executives look for in a coach, a good executive coach does not need to have the exact background or experience as the CEO , but a familiarity will help him or her better understand the CEO thinking and needs. More importantly, the CEO coach needs skills, the CEO either does not have and wants to attain, or ones that can help strength the CEO opportunity areas.

In general, what leaders want from coaching. Coaching empowers leaders to do expectional work. Coaches establish and advantageous relationship that uncovers hidden strengths and weaknesses within the leader. Goals will be created to enable leaders to indicate their weaknesses and track their progress. What is better up coaching? Coaching training can address the CEO must pervasive organizational challenges with the organizational unique combination of coaching to achieve a growth approach to mental fitness and organizational health more effectively.

Have a CEO coach is similar to the executive or leadership coach , but with the added responsibility of working with the CEO who is working in the firm. So, coach can attempt to help the CEO potentially make the most significant difference in the company's success and lives and careers of those who work for the company. Hence, any organizations can provide executive or leadership coach to improve the CEO's leadership skills by these 5 caoching styles, such as below:

Democratic coaching, this method gives the term freedom and accountability, with the coach stepping in only when meeded to keep the process going, or authoritarian coaching, holistic coaching, authcratic coaching and vision coaching training methods, for example, entrepreneur coach can help new and existing business owners with any number of tacks to foster their entrepreneurship. In fact, entrepreneurship is not about having a business, but about having an entrepreneurship mind. So, CEOs can be trained from any style of coach in order to improve their management and leadership skills to more proficient in their organizations. So, it means that coaching training is one kind of effective training method to improve CEO's performance in any organizations.

On conclusion, I believe that it is only one kind of training method to train any organizational CEOs to become proficient CEO, it is coaching training method, instead of providing general educational level to the CEO for leadership knowledge, because coach training is one kind of organizational leadership practice training, it can satisfy any CEOs to raise leadership effort effectively to help the CEO to manage his/her organization easily.

Amazon leader / founder owns new business foundation strategies mind

Any new business founders, they ought hope their new businesses can run long time. The question concerns that how they can help their new businesses run long time, e.g. at least above 5 years . I shall attempt to apply behavioral economic theory to solve this common challenge as below.

Keeping a new business is in difficult economic time is challenging. Every new business is different and each carries its own risks and rewards in behavioral economy view. These sifferences cause some new business founders attempt to copy another similar new business founder strategy. Still, these are save general strategies business owners can follow to help themselves new businesses to copy another similar new business strategy succeeds in possible.

However, some of these copying another new business founder strategy's owner still feels their sopying another new business founder strategy may help them to succeed. So, they still have risk or they may encounter failure, if the another similar new business founder strategy can not be suitable to be adopted to themselves new business similiarly. So, it explains that why some new business founder can not keep their new businesses can run long time, because they feel the other similar new business founders their strategies can help to develop their new businesses succeed together. But in fact, there are may new business founders their copying strategy decisions are wrong. They feel their copying another new business founder's strategy can help them to develop in success. So, their new business strategies ought also help their new businesses to develop or grow long time. However, there are many new businesses can not run above 5 years, due to there new business founders choose the wrong copying strategies from another similar new business or old business competitors.

New business founder needs to look at the big picture, it means that long term customer behavior change picture. Because consumer behaviors must often change suddenly in any time. So, any new business founder ought attempt to discover whether what their product buyers behaviors will change after 5 years, even 10 years in predicting. People have a tendency to attack the most obvious immediate prodblems without hesitation. That's understandable and might make good business sense in some suitations. However, it is also advisable to step back and look at the big picture to see what is still working and what might need changing.

Its an opportunity to better comprehend the size and the scope of exciting problem and further understand your new firm's decision model, determining how its strengths and weaknesses come into play. What is a business model? The term business model refers to a firm's plan making a profit. It identified target market, e.g. which is age customer group, where is the geographical sale place, and anticipated expenses. However, business models are improtant for both new and established businesses. They help new developing companies attract investment, recruit talent, and motivate management and staff. Establishing businesses should reguarly update their business plans or they will fail to anticipate trends and channelgens. So, business models both levers are pricing and costs. It is a high -level plan for profitably , a business in a specific marketplace. A primary component of the business model is the value proposition. This is a description of the goods or services that a new business offers and why they are desirable to customers, ideally stated in a may that differentiates the products or services from its competitors.

A new business enterprise's business model should also cover projected startup costs and financing sources, the target customer base for the new business, marketing strat egy , a review of the competition, and projections of revenue ans expenses. It may also define opportunities in which the new business can partner with other established companies, e.g. the new business model for an advertising business may identify benefits from an arrangement for referrals to and from a printing company. So, successful new businesses need have good business models that allow them to fulfill client needs at a competitive price and a sustainable cost. Over time, many new businesses revise their business models from time to time to reflect changing business environments and market demands. However, the business model may not tell new business founder everything about a company's prospects, but the investor who underatands the business model can make better sense of the financial data.

New business founder also needs to consider organizational internal matter, e.g. suppose a new business founder discovers that two employees are making mistakes with inventory that cause certain supplies to be overstocked or understocked. When a initial reaction might be to fine those employees. It might be wiser to examine whether the manager who hires and supervises them properly trained time. If the manager is to blame, that person could be fired, but this might not be the best solution. If the manager's relationship with client have a history of bringing in repeat business and substantial revenue. They are likely someone, you would want to keep. However, retraining might be a better alternative than termination. Infact, by thoroughly reviewing the strengths and weaknesses of the employees, the owner is looking at the issue from a top-down perspective, reducing or eliminating the chance that the problem will occur when avoiding a change that could adversely impact future sales. Hence, a similar kind on analysing how youe new products or services fit into the marketplace in your new business beginning stage, how the economic crisis has affected your customers and suppliers and all the other key aspects of your new business. You need to know how well your new business model fits the current environment and forecast what various alternative scenarios of the future mighr mean for it.

For one interesting new business behavioral economic view to recuriting new employees view example, any new business owners or large corporational founders tend to be either wise or follish when they hire the least expensive workers sometimes, the productivitiy of these workers may be suspect. Hiring one worker who costs 20% more than the average workers, but works 40% more effectively make of crisis. By seeking resumes and interviews from new applicants. New business founders can make change to new staff when needed to increase efficiency. So, how to choose any department new staff recruitment , it is very important to influence the new business furher develops for long time. Don't sacrifice quickly, keeping a handle on costs is crucial in tough times. Owners need to stay on the offensive and get employees on board with changes that are being made. However, any one new business founders need have good sense to predict when their new products or new services need to be changed in order to adopt customer behavioral changing environment.

ON conclusion, so any new business founders hipe to keep themselves new businesses can run long time, they must need to know " consumer behavioral psychology" how and why to cause their change in order to implement the most effective new business strategy to improve their sale methods to achieve the most satisfactory level to their potential clients' purchase needs or service needs in long time. Consequently , their new businesses ought keep long running time in this old and new product / service competitor market.

Avoiding new business low value method

Internet marketing promotion strategy

Any new business founders do not hope their new business market worth falls rapidly or share price falls rapidly. What facors may cause new business share price or market value falls rapidly? What methods to help new businesses keep the same share price is stable long time or market value can be kept in the stable worth, even share price can be influenced the rise or market value can be influenced to rise? I shall attempt to explain some useful methods, they may influence new business market value to avoid falls down, or share price can rise more easily.

TO avoid new business low value, these ways may be used in developing new business. They may include: Knowing what your client individual actual need, e.g. rice cooker product, cooking rice function must need , but avoiding to spend long time or short time to cook rice rapidly, big rice cooker size to cook more rice, long time keeping rice warm function. All these factors may influence rice cook buyers individual choices. Some housewives also need the kind of rice cooker has above all these functions, before they make rice cooker purchase decision. So, knowing what your clients real needs, they are essential to product manufacturers, offering great customer service, e.g. repair sevice, after sale enquiry (following up equiry to the client, nurture existing customers and look for new sale need opportunities, use social media attending networking events, give back to your community, measure what works and refine your approach as you go. So, if the business founder can manage sale and customer, service both teams to achieve the best service performance to let clients feel , then they may persuade many clients continue choose to continue to buy their products more easily. So, salespeople and customer service staffs training method may be one main influential factor to help the organization to raise market share value. If the organization can design useful training course to raise its salespeople sale skills or customer service staff service performance.

Which businesses can be started with less investment? For India country example, these low investment business wil be helped the new business founder to earn the most profitable, e.g. dropshipping is one of the most successful business in India, because in India, even global there are many tall offices need dropshipping equipment to transport window cleaners to clean tall office windows. So, dropshipping cleaners'window ability may influence India deopshipping firm share price changes to rise up or fall down. Because India dropshiping companies window clean workers window cleaning performance can bring property management company clients to make dropshipping window cleaning service provider choice.

I mean that India dropshipping office window or global dropshipping office windoe clean service providers, their new business market value or high stable shar e price value, their value is depended on whether their dropshipping window cleaners clearning window skills or window clean priperty management clients' window clean needs. So, for dropshipping office cleaning service provider case example, how to improve dropshipping office window cleaning workers' windows clean ability, it is only one influential method to influence dropshipping office window cleaning firm service providers their shares prices whether they can keep stable market position, even share price can be risen easily. How to improve dropshipping office cleaning workers' office windows cleaning skill is very important to influence dropshipping office window service providers' shares prices are at the stable high price level. For courier company example, how to improve courier service performance, e.g. shorten the document / goods delivery time between the document or goods sender countey and the document or goods country receiver, when the document or goods can not bt delayed to send to the overseas receiver whose home from the another country sender, e.g. From US sends the document to China in common general couriers need 5 days , but the US courier company can send the documents to China, it only needs 2 days delivery time. Then, it may attract many China goods or documents receiving clients to choose the US courier service providers. So, how to keep the US new courier servier providers, or market value or share price value, it depends on whether it may help its overseas documents or goods receiving clients to reduce how long time to let they can receive the documents or goods from US. So, the avoiding delivery time delay is the import factor to influence courier goods / documents delivery service provider new business market value.

Moreover, any one new business founder may promote themselves new business in a low budger. These methods may include: posting amazing content on your new firm's blog, creating a google my new business account, building a free or cheap email list, contributing an article to an industry magazine, attending local networking events, co-sponsor a contest. So, internet channel may help any new business to build a rapid promotion network to let many people to know that this kind of new business exist, it aims to let potential clients number increases in short time in possible. So, internet may be a kind of new promotion tool to help any new business market share value rises in short time, the internet strategies for effective promotion of new business, e.g. creating a website, geting listed on google, advertise on facebook, email customers, and potential customers use google Adwords, paid media advertising , social networks and viral marketing, internet marketing, email marketing, direct selling, point-of-purchase marketing, co-branding, cause marketing , conversational marketing .

All of these internet promotion methods may help new businesses to rise market growth value or share price in short time. Hence, internet promotion method is preferable to choose to compare general TV, advertising , newspaper, advertising, magazine advertising, radio adverting, sales promotion, general selling, publicity promotion methods because internet promotion marketing is one kind method, it may bring the new product or new service providing message to let many potential clients to know from the new firm's email ot website in short time rapidly. It is one kind of new global promotion method for any kinds of new busienss development in beginnning.

So, any new business founder may prefer to choose internet promotion method in orde to let many potential cients to know their new products ot services are existence. So, attracting new customers method may use social media optimizing the new business founder's social media account, improves website and engage with loyal customers, give branded gifts , referral discounts, social media contects, and giveways, sending email to survey customers, researching your competitors and finding out who their customers are, target email advertisement, smart social media, responding to every email, tweet , facebook comment, e-publish user reviews from internet media channel. All of these new internet promotion strategies may help any kinds of new businesses to promote their products or services to let clients to know in short time rapidly from internet media promotion channel. The internet promotion marketing strategy is one kind of low cost promotion strategy, low cost marketing stragtegy for startups, cheapest new businesses to start and focused low cost strategy company.

Why internet promoting strategy can help new business to avoid low market value? The reason is simple, because when the new firm sends one email message to let any one country's email user to know. This new product or new service email message can let global any one emaill user to know that this new product or new service business is existence in market. When the email user receives the new business' semail message, it may keep in itself email box. So, the email user won't lose the one email message and it can often remember this new product, or new service provider is existence in market. So, any one email user when he/she receives new product or service provider 's email, this email is such as the private advertisment between the new business founder and the email receiver. So, email advertising is different general product magazine, TV, newspaper , radio advertisment. It is public promotion advertisement.

Hence, the feeling of private advertisment, it is the most influential factor to persuade potential consumers to choose to buy the new product or use the new service from this provider, because when the potiential client receives the email promotion from the new product/ new service provider, he /she may feel surpise to raise interest to click the advertismeent email to see whether what benefits can be given ti him/her if he /she chooses to buy the new product or use the new service. If one say, the new product / new service provider can send about 1,000 promotion email to let global 1,000 different countries possible potential clients to know its new product/new service is existence in market. Consequently, this email network promotion method can help this new business founder to advertise his / her new product /new service to let global 1,000 email users to know in one day. When, its potential clients number increases, it implies that its new product or new service market existence value will be possible to influence to rise up. Hence, emial promotion method may influence any kinds of new businesses to rise market value in short time in possible, because when may email receivers become to the new business potential clients. Consequently, the new business's share price may be influenced to riase up then its market value may also be follow to influence to rise up. So, it seems that email promotion method may be nowadays a kind of the most effective promotion method to help any new businesses to raise up share price or raise market value in short time.

Becoming talent human how influences social changes

Nowadays we tend to think about social and digital technology more from a personal or consumer perspective than their business or professional applications, but as the Digital Era continues to progress, many of technology's most profound impacts are likely to be in the world of work. In addition to changes in product and business development, knowledge management, data analysis, and other operational processes, transforming talent management will be a key priority for organizations striving to be employers of choice.

● Digital technology encourages to create talent human

Why does digitial technology encourage to create talent human or excite human to learn new things ? The human capital implications of social and digital technologies impact virtually everyone, regardless of the type of organization they work for, their profession, their functional area, or their career stage. That means that the talent management functions in all organizations, as well as the professionals who staff and lead them, have a critical role to play in ensuring the efficient and effective transition and transformation from Industrial Era models and processes to their Digital Era upgrades.

It's no surprise that talent management has already become more "high tech." Many employment related activities have been digitized, and there has been a corresponding increase in employee self-service. It's important to remember, however, that digitization is not the same thing as digital engagement, and that the rise of "high tech" solutions doesn't necessitate the loss of a "high touch" approach to managing an organization's human assets. Transforming talent management requires digitization, to be sure, but it also involves leveraging social and digital technologies in ways that promote and enhance communication, collaboration, and engagement - not just between an employee and the organization, but between and among employees themselves.

Talent Acquisition

The logical place to start when talking about the impact of social and digital technologies on talent management is talent acquisition, where the greatest advances have been made. Anyone who has searched and applied for jobs in the past 10 years is very familiar with how technology has transformed the application process, which in most organizations (and virtually all large ones) is now almost completely digitized and automated. However, there are other ways in which social and digital technologies are impacting talent acquisition that may not be as well-known or commonly understood. Social media sites in particular (such as Facebook, YouTube, and Pinterest) are a great way to promote an employer's brand and offer realistic previews of work life, people and culture in organizations. Online games and simulations can also be used to get a sense of what working for an organization would be like, and give organizations themselves an opportunity to determine if a prospective candidate would be a good cultural fit and potentially successful.

On organizational working environment aspect, some employers are recognizing the value of digital alumni networks or communities to maintain strong relationships with former employees. One of the primary motivations for doing this is that the employees may return one day and/ or make referrals to or from their personal and professional networks. Similarly, talent networks enable organizations to establish and maintain relationships with professionals in key areas like IT and engineering, even when there isn't a current opportunity to have those folks be a part of the organization. Moreover, social media can bring positive influence to impact organizations to encourge employees to attempt or feel needs to learning new things for their tasks needs. Due to social media has actually transformed every stage of the recruiting process in significant ways - so much so that the traditional recruiting funnel can be recast in "social" terms. At the top of the funnel are activities like social advertising (i.e., placing job ads on social networks like Facebook), social sourcing (i.e., searching for candidates who meet certain criteria on networks like LinkedIn), and social referrals (i.e., having current employees share position openings with their online personal and professional networks). And at the bottom of the funnel is social screening (i.e., reviewing a candidate's public activity in social networks to identify potential hiring risks).

● Learning management is needed to feel needs to any organizations

Learning management is probably the another most advanced area when it comes to adopting and adapting to new technologies. As with recruiting and other processes, the initial advances are in the area of digitization, with social software applications evolving next. One of the obvious digital impacts is the increased use of elearning and online learning platforms with self-paced study. There are also countless instructional videos on the web, both free and fee-based, that address a virtually unlimited range of topics. And we can't forget MOOCs - massive, open, online courses - which have proliferated in the past couple of years. Finally, many organizations have also started to leverage tablets and other mobile devices for learning, as well as using simulations and games to help employees develop specific skills. In addition to offering training through a variety of multimedia channels, organizations are increasingly using a range of digital tools for assessing employees‘ skills. They're also allowing employees to play an enhanced role in identifying their key skill sets and training needs, and can even have them create their own learning and development plans. Allowing employees to take a more active role in their own learning and skills management enables organizations to develop and maintain a more complete and accurate knowledge and skills database, which in turn enables them to maximize the value of the workforce in which they've already invested.

1. Formal learning management systems and platforms are also beginning to incorporate social technologies in a variety of ways. Promoting connections and interactions among participants, as well as with the instructor, can enhance the learning experience both during and after a course. Creating course-based cohorts that allow people to continue to interact with each other via a digital community - even when their shared learning experience is face-to-face - can promote both knowledge transfer and retention, in addition to increasing commitment and engagement through interpersonal connections.

2. Informal learning - which is now also referred to as social learning - is greatly enhanced by social technologies as well. In fact, this is probably the greatest opportunity and area of growth for organizations of all types and sizes. Through private social networks, intranets and other internal platforms that have incorporated social technology elements, organizations are better able to facilitate employee learning as they perform their job duties and complete work activities. Along with the networks themselves, features like advanced search, identified subject matter experts, digital communities of practice, wikis and more enable employees to access and learn from colleagues who are not just next door or down the hall, but even in another city, state or country!

As organizations move forward with leveraging technology to enhance learning initiatives, it will become increasingly important for them to address issues related to digital literacy and digital competencies. For the past several decades we've generally taken what I refer to as an LIY, or

Learn It Yourself, approach to digital knowledge and skills. Although organizations may invest in teaching someone how to use a specific application related to their job, they make virtually no investment in helping individuals learn how to use general digital tools like Microsoft Office and even email. Left to their own devices, most people - and I include myself in this group- are much less efficient and effective at using these tools than they could or should be. As our tools get even more sophisticated, we need the foundational knowledge and skills to be able to use them well - and this foundation should probably be provided via more formal training. In other words, many people need to be "taught how to learn" in the Digital Era. If organizations aren't going to provide the formal training workers need to do that, it's probably in an individual's best interests to pursue those kinds of development opportunities on their own.

● What are social influences on human behavior when talent human number increases?

Social Influences on Human Behavior Because human beings are social and learn from observation rather than depending entirely on instinct, almost all aspects of human psychology and behavior are socially influenced. Languages, modes of dress, gender roles and avoided taboos are all agreed upon at a group level and form the basis of culture. What are the characteristics of social change? Small-scale and short-term changes are characteristic of human societies, because customs and norms change, new techniques and technologies are invented, environmental changes spur new adaptations, and conflicts result in redistributions of power. This universal human potential for social change has a biological basis.

This universal human potential for social change has a biological basis. It is rooted in the flexibility and adaptability of the human species—the near absence of biologically fixed action patterns (instincts) on the one hand and the enormous capacity for learning, symbolizing, and creating on the other hand. Because human beings are social and learn from observation rather than depending entirely on instinct, almost all aspects of human psychology and behavior are socially influenced. Languages, modes of dress, gender roles and avoided taboos are all agreed upon at a group level and form the basis of culture.

On conclusion, when one country can create many talent human, e.g. students, workers. Then they can bring more positive attribution to help or assist themselve country to develop rapidly. Consequently, the country's economic growth speed will be rapid. So, I believe that it has close relationship between economy growth and talent human number to any countries in nowadays societies.

Amazon ecommerce founder successful factors

Nowadays, we are encountering digital age period. Many customers choose to apply websites and mobile apps to buy their products. So,many merchants also choose online e-commerce channel to build online purchase platform to raise its competitive ability in online e-commerce market.
So, customers can apply mobile apps or merchants themselves websites to buy any products in any time and any places conveniently.

However, in electronic (ecommerce) industry, Amazon is the global largest online retail delivery provider. It sells different kinds of products from internet, e.g. books, electronic products, music, movie Cd, DVD, magazine, garden and homw useful tools, children toys, computer,
software, even, cars from online channel to global online shoppers. So, there ate many online buyers choose to buy any products from Amazon web services middle channel. In fact, its products prices can be low, a wide selection, ease website use and convenience to meet all of its customers' needs in one virtual store. So, it can still own the high share market in online retail delivery service industry. I shall indicate that Amazon successful factors as below:

It has a clear aim or mission. It's mission is to be Earth's most customer-centric company , where people can find and discover anything , they want to buy online. So, it will gather data ro analyze whether why and how customers will select to buy the kinds of products from internet. Why do its customers forget to choose to buy any products from shops and they choose Amazon to buy their preferable products from online? I believe that it has these characteristics to attract them such as:

It can provide return after sale services within the reasonable time when the customer feels that he does not need to use the product or feels unsatisfactory to the product, then he can return the product to Amazon and Amazon must refund money to him as well as it can provide free charge of grocery delivery service to every customer home. Instead of these attractive service, it also began to enter publishing market. In July 2002, Amazon started offering services to website developers, marketings its kindle
product, aimed at capturing the publishing market for digital books, then it can sell paper books from online channel. It aims to let readers can choose

to either download to read ebooks from its website or borrow ebooks to read from its electronic library or paid visa to print paper books to deliver their paper books to their homes conveniently. So, readers feel that they can enter Amazon publish website to choose any interesting books to buy or borrow to read , they do not need to go to book stores or libraries.It is Amazon's competitive and attractive and unique strengths to build its customers on this online retail market.

On its organization mamagement aspect, it has one effective and efficient management strategy. It has a good CEO manages his e-commerce business. Bezos has been chairman of the board of Amazon , since he founded the compnay in 1994. Amazon has a limitless stock on hand at all time, it enables Amazon to collect high margins when providing low prices, and lets customers to feel consideration every second. So, when the customers feel any enquiries, his customer service team members can apply its intra website email message channel to send email to answer his enquiries when they enter Amazon intra-website email message channel to send email to ask them any enquiries and they can reponse their enquiries immediately in any time or they can phone to Amazon customer service hotline to enquiry them directly. So, its customer service staffs can answer their enquiries either by phone or email communication in short time rapidly. They

won't delay to response their enquiries to avoid they feel worry or unhappy or complain. So, efficient customer service performance can help it to satisy its customer service needs and build good customer service relationship and repeat purchase chance will increase also.

However, global online retail marketplace is expanding rapidly. So, Amazon will have many new or potential online retail competitors, if it neglected to improve its strategy to adapt every online customer individual need or purchase taste, due to customers‘ purchase need will often change, such as price demand, product delivery time demand, customer service demand etc. spects. So, Amazon needs have different market strategies in different time in order to attract customers' choices in different economic environment, if it hoped to keep online retail leader position.

For example, how to encourage or raise customers‘ online purchase desires when economy is poor environment. When many people loss jobs, due to many businesses liquidate and loss many clients. So, they need to dismiss many staffs. Then, in society, many families will reduce their consumption desires because their parents loss jobs. So, Amazon needs improve its strategies on above different aspects in order to attract or persuade them to spend much time on internet shopping activities.

Instead of organization management aspect and customer service aspect, the other critical success factors for developing an e-business strategy to Amazon, they may include: How to apply network technology to keep long term close relationship between Amazon its e-commerce organization, online customers, partners, stakeholders and product suppliers. It is very important to influence Amazon's success, because Amazon , such as the e-business delivery service provider will loss the kinds of product sale chance, if the product providers dislike its sale delivery service or feel satisfactory to
its sale strategy or promotion methods or when its online sale delivery service can not let its customers feel its online sale delivery to be satisfactory. Hence, online sale delivery service performance is very important to influence Amazon's success.

Moreover, in its business model, Amazon.com also needs have these key succes factors. They may include: Building strong brand name location, because when online shoppers can remember its brand or loyalty easily when it is famous. Then they won't choose any online retail delivery providers to replace its delivery service easily. So, building one famous and confident online retail delivery service provider loyalty or brand , it can help Amazon to influence many online shoppers have confidence to choose its online retail delivery service in the first time. When they can often remember Amazon and feel it is only one online retail delivery serivce provider, the repeat online purchase chance will also increase, due to they only choose to click on its website to choose any kinds of products to buy more easily.

So, Amazon.com's marketing strategy is needed to design to strengthen the Amazon brand name, increases customer traffic to the Amazon.com web sites, builds customer loyalty, encourages repeat purchases or attracts them to clicks on its websites again to develop incremental products and retail delivery service revenue opportunities. Also, how to design its efficient products delivery value to let product buyers to feel. This factor is important to influence Amazon's products delviery service in success in online
retail delivery service market. Moreover, Amazon publish service also needs to build attractive reading feeling to let readers to satisfy its reading provision needs to replace book stores or libraries.

On conclusion, Amazon ought need to continue to change its marketing strategies in order to adapt to its online retail product delivery service changes as well as lets they believe that it is one online customer care product delviery service provider to comapre other online retail delivery service providers, if it hopes to keep its online retail relviery provider top leader position in this e-commece market.

Skills technological improvement strategy
soft and hard skill to Amazon organization

Skills shortages on developing country market
Nowadays , future global job market competition will be trended serious. Any employers will expert their employees own different skills to know how to do their jobs efficiently and effectively and easily. So, future any organization employees ought considerate how to learn different kinds of skills or knowledges in order to prepare to satisfy their future employers' different new tasks needs. However, if future any new skillful needs or demands will be raised to future employers' demands. It brings these questions: what skills do global any organization employees need own in general? How to improve or raise employees themselves skills more easily and efficiently? What will happen if future employees do not learn new knowledge to improve or raise themselves skills? Why is learning any new skillful knowledge important ? What will be the possible negative and /or positive consequence if future the organizations do not need their employees to learn any new kinds of skillful knowledge?
Future developing countries need to develop their economy, so they need to employ many employees who own technical skills and /or soft skills. What kind of technical skills and/or soft skills , the developing countries' employees who will need to order to raise competiton in local job market ? I shall indicate the developing country China example. China is one developing country, employers will need different kinds of skillful labors to assist them to develop their businesses. However, China employers will face skillful labour shortage challenge. Although Chinese young age population is high, but many of them do not to be encouraged to learn enough skillful knowledge to fill future new skillful positions. So, the fast speed of training will be important to influence China supply and demand labour market to be more accurately as well as future China's the quality of labour demand number will be influenced to be raised after they have enough training to learn new skills.
How to solve future China skillful shortage of labour? Firstly, nowadays,

China employers need to teach their employees to learn how to use and how to operate robotic skills in China's factories. AI robotic has been early developing, so they need to prepare to learn robotic management and operating technical and soft skills in order to satisfy future China factory automation industry development.

China is one world's factory for low-end products to high quality information products, high end technology and services. So, China will need many high skilled workers to assist manufacturers to manufacture many different kinds of products to export or local sale. Moreover, robotic manufacturing skillful workers will also need because robotic will be accepted to assist manual workers to work in China's any factories. This has led to greater demand for labour wirh upgraded skills and competence. So, it seems that China's orkers need to lern any high technological manufacturing knowledge, e.g. learning how to co-operate with robotics to raise productive efficiencies, which will be future many China's manufacturers‘ skills need intention.

So, when any one of China manufacturer invests robotics to work in its factory . Then, the China manufacturer's labours ought need to know how to co-operate with th robotics to raise productivities and efficiencies. Moreover, these China service industries, e.g. IT, software, accounting, finance, marketing and customer service management, e.g. waiter, property security, shopping center customer service etc. service occupations. In the future, robotics can also used to participate any one of these service industries' part of tasks in order to raise service performance. So, any one of these service industries‘ employees need to learn how to operate with robotics in order to achieve the most excellent servvice performances to satisfy consumers' needs. So, China service industries labours ought need to learn how to co-operate or manage service natural robotics to work together more efficiently because future China manufacturers will prefer to employ the labours who know how to co-operate and manage and control any service natural robotics more easily and efficiently in order to achieve the most excellent service performance to satisfy customers needs.

Hence, it seems that China manufacturing and service workers need to spend time and effort to learn how to co-operate with manufacturing natural robotics to manufacture any products in factories efficiently or deliver any cargos in warehouses more efficiently or serve customers to let them to feel excellent service performance in restaurants or shopping centers or properties or offices receiption counters. Then, when their China employers apply robotics to participate to work in factories, restaurants, shopping centers, cinemas, offices or properties reception etc. different working places . These low skillful labours will be dismissed easily, due to robotics can replace them to manufacture any products or provide services to satisfy clients' needs in order to let them to fell robotics' performances are more excellent to compare human service labours or their productive efficiencies are more effort to compare workers. So, future China workers need to learn how to cooperate or manage or contol with robotics to work more efficiently, if they do not expect to be dismissed easily.

Future global skillful labor
soft knowledge skill need

In the future several occupations have been identified as the most frequent movers between all labour market states. The elementary occupations include: waiters, bar staffs, clearners, catering assistants, construction and security service workers, care workers, sales assistants and general clerks etc. So, the low educational level workers can learn these soft wkills to raise whose professional workering level to prepare to do these above positions in global elementary occupation job market.

The changes of employer were most frequent for IT programmers, doctors, electricians, carpenters, skilled workers in global labour market. These skilled occupations will have manpower shortage supply challenge, due to either people feel the educatonal level is under low. So, there has no many people have interest to know these knowledg to prepare their elementary careers. So, these kinds of low skilled occupations will have not enough human power supply to global labour job market also, the high skilled or educational job support.

Moreover, the high skilled occupations also encounter labour shortage issue. The skills in short supply related to experienced canadidates e.g. five years or more. For example, pharmaceutical , biogharma and food innovation industries. The occupational shortage roles include: Chemists, analytical scientists, product formulation, analytical development for roles in biopharma, quality control analyst includes pharmaco-vigilance, i.e. drug safety roles. The demand for engineering industry aspect which will aos increase the labour shortage includes process and design (research and development, quality control, automation, lean processes) are skillful labours need to help employers to achieve these intentions. They may include raising competitiveness, boosting productivity and skills availability. So, if future these above any one of occupation labours can not achieve these benefits to satisfy their employers' needs. Then, his/her average weekly or hourly wages will be reduced. It means that the unskilled labour under skilled labour wage can not increased more easily, even they own many year working experiences in any one of above these occupations. If the employer feels the labour is unskilled or below skilled level for any one of these occupations in these any one industry aspect, e.g. wholesale and retal , human health, education, accomodaton and food , construction, professional activities, financial service , public administration, and defence, transportation etc. occupations. Then, these industries' unskilled or below skilled level workers' salaries will be lower level to compare the higher skilled workers who work in any one of these industries.

The reason why future employers need to employ skilled labours. One explanation for slow recovery in demand in negative impact on investment is a prolonged period of high unemployment. This is led to job weekers left labour market or became unemployable due. So, future low skillful level will be one important factor to cause unemployment in society as well as nowadays labours ought need consider whether their skills are needed to improve in order to avoid future competition in job market.

2.1 Why do future labours need to learn worldwide readiness skills

Future employers need employees own worldwide readiness skills, such as reading , writing and arithmatic. Why do employees need worldwide readiness skills? In the future, high economic growth countries need high wage positions, high opportunity jobs which need a large number of skills required of job candidates of these positions " job readinss" and not " job training" , which support developments of these importance and widely desired skills won't only support the success to high-opportunity positions, but also be developed for future success in the competitive global economy. Because real-time business intelligence is needed for the talent marketplace to employ talent employees. So, it explains that it will have many future employers hope to employ owning readiness skillful employees to help them to develop their businesse intelligently. Hence, present employees ought need to hard to train readiness skills to prepare whose future employers' job requirements in the future competitive global job market.

2.2 Why these occupations need readiness skills

In the future these occupations will need to raise readiness skills. For example, mathematical science, teachers (post-secondary), management analysts, computer and information systems, managers, first-line supervisors of construction traders, solar photovoltaic installers. All of these occupations , employers need staffs to own good readiness analytic ability to help them to do more accurate real-time business intelligent decisions. The representative occupations include oral and written communication skills, project management, teamwork, marketing and creativity . Moreover, they need to own specific technology skill, deep science and math or even most business skills as well as these skills are "soft" skills more than hard skills. These kinds of occupation employees need own cooperative effort, creativity, problem solving, detail orientation and integrity personal characteristics, which are relevant across all knowledge and domains.

Therefore, in the future, science, technology,engineering and mathematics relevant occupations need to own more readniness and analytic skills more than othe kinds of occupations. Because these organizations need those professionals on knowledge acquisition, literacy analysis, synthesis and critical thinking skills that will impact their organizations to bring more critical thinking beneficial team culture. These occupational top skills will include oral and written communication skills, project management skill, team oriented skill, marketing and creativity skills, problem solving skill, detail oriented skill, self-motivated skills, management and analytical skills, coaching skill, business process modeling skills, work independent skill, strong leadership skills, management experience and business requirements gathering. All of these skills which will be future employers who need to employ these kinds employees who own these skills in preference. Also, all of these skills concentrate on soft skills more than hard skills. It seems that when above occupational applicants who own any one of thes skills, evn more than one skills. Then, he/she will have more chance to be selected to employ. Also occupation specific skills requirements are more needed to compare cross-functional skills for above of any one occupation. Because the high concentraton of cross-functional skills require " job readiness" and not " job training" for success, e.g. communicaton, integraton and presentation skills, entrepreneurialism and related skills, microsoft office software skills.

Of particular interest is communication, integration and presentation skills. These skills include ability to seek, evaluate and examine information and data create a reasoned position, present findings and make a case for or advocate for position. So, these skills are very important and they can help future applicants who expect to win any kinds of these positions easily.

However, the hard skills can help these applicants to be more successful to win any kinds of these positions when they own these hard skills, e.g. microsoft offic, powerpoint, excel , word, microsoft project etc. softwares.
In conclusion, the global economy is dynamic and many of the skills required for positons in the future will need good technologies and work practices to be developed. The number of skills required t be successful in the jobs forecast to be most in demand in the future is growing. So, it explains that why future any one of these occupations which will need soft skills more than hard skills, due to organizations like to employ the employees who own managerial and analytical effort more than hard skills productive effort to assist their organizations to develop more easily.

2.3 Data -analysis skill needs

In the future, most organizations will have a number of jobs that include data analysis. Economists and labor market forecasters predict occupations need data analytical skill will need much. In addition, fast technological development means th types of technologies and applications workers in this field will need to be familiar with data analytical skill rapidly. It seems that data analytical jobs will have new job opportunity to employees with in-demand skills in future global labor market.
Why and how do employers demand for data analysis skills? Data analysis skills mean the ability to gather, analyze and draw practical conclusions from data as well as communicate data findings to others. The occupations include: data analyst, data scientist, statistician, market research analyst, financial analyst,research manager. In business career, many employers expect to employ statisticans, operations researh analysts, market research analysts and marketing specialists to assist their organizations to gather useful data from market in order to analyze and draw practical conclusions and finding the best solutions or methods to win their competitors.
Therefore, these data analysis jobs will have much need. Large size organizations with 500 or more employees were more likely than small or medium size organizations with 25 to 499 employees to plan hired data analysis positons in the future. For example, human source department will use big data to help make strategic decisions. How HR uses big data . HR will use big data for sourcing, recruitment, or selection, identifying causes of turnover and/or employee retention strategies or trends, managing talent and performance. Why organizations do not use big data. It is possible that they lack of knowledg expertise, the majority of organizatons will

have data analysis positions within accounting and finance department, human resources department, business and administration department, information technology department, marketing, advertising and sales department, supply chain and operations department, research and development department, customer service department and other departments. So, future data analysis skill will need to used in different organizational departments.

However, publicly and privately owned for-profit organizations were more likely than government organizations to have data analysis positions in the marketing, advertising and sales function. Also, data analysis skills are required to different levels in any organizations , such as entry level, non-management / individual contributor level, mid-level management level, seniot management or executive level. The analyst, research analyst, market research analyst, scientist-based titles include: data scientists , research scientist, scientist, other descriptive titles include researcher, statistician, mathematician and other . So, data analysis positions will have many different skills to be selected to any one data analysis professional. For example, the data analysis professional can select either to learn the ability to interpret and communicate data analysis results skill or to learn how gathering or analyzing data skill. So, data analysis skill is not onlyone skill, it is more than one skill to let any one employee to select to learn.

Why do organizations need data analysis professionals? On workforce planning aspect, organizatons expect to let strategic direction and content of workforce needed for future business objectives easier, analyzing workforce: supply analysis, demand analsis and gap analysis more earier, developing action plan : recruiting and training plans to deal with gaps more easier, implementing action plan, monitoring, evaluating and revising plan more easier. So, organizations expect the data analysis professionsla can help them to solve these challenges, such as using of advanced technology solutons to integrate disparate planning sources; data availability and format; accessing to and understanding of the organization's data and analytics, developing business case to gain support from senior management and collaboration among HR staff, managers and executive easier. Future industries need data analysis professionals may include manufacturing health care and social assistance, scientific and technical service, finance and insurance, educational services , government agencies, retail trade, transportation and warehousing, construction, utilities, accomodation, and food services, waste management and remediation services, entetainment, and creation, real estate and rental and leasing , repair and maintenance, agriculture, forestry, fishing and hunting, personal and laundry services etc.

In conclusion, data analysis job need explains why future readiness and data analytical skills will be popular needed in globl labour market , due to these both skills are labour shortage and employers will need employees own big data readiness and data analytical both skills in order to win whose competitors more easier.

2.4 What are regional dynamic skills
of global labour market demand

Businessmen expect to improve better economic environment, they will prefer to recruit the most sought after skills of intelligent employees to bring positive beneficial impact to organizations. However, technology and digisation has had a significant influence on workers. Future globalization will trend digital economic development. Hence, it will influence workers' skills to be changed also. In fact, not all changes are positive because some workers will possible lose jobs, either due to new technology replaces their jobs or they lack enough effort to improve their skills in global digital economic labour market environment.

It brings this question: What are regional dynamic skills need whn digital busines environment is growing. In fact, organizations will continue to deal with skills shortages, labour markets across the global are continually changing. so, more employers and workers will need to adopt innovate working pattern, e.g. on call jobs, freelance jobs will grow popularly. The greater flexibility afforded to employ regardly.

Finally, digitalisation includes artificial intelligence, big data , online platforms. All these new technology will influence future employees how to worker. For example, they can apply online platform to work at home conveniently. So, they do not need to go to offices. They can finish their jobs and send to their employers by email easily. This kinds of job pattern can raise efficiencies and employers do not need go to offices often.

An important implication of innovating working which needs the employees who own digital skills in order to serve organizations more efficiently. So, employers are increasingly able to access demographics that were hitherto less active in labour markets. For example, future more women are joining the labour market because part time and self employment opportunities make it easier. This kinds of job pattern can raise efficiencies and employees do not need go to offices often.

An important implication of innovating working which needs the employees who own digital skills in order to serve organizations more efficiently. So, employers are increasingly able to access demographic that were hitherto less active in labour markets. For example, future more women are joining the labour market because part time and self employment opportunities make it easier to manage family with work life. So, digital skilling needs will cause many women lose jobs in possible. If the women lack digital job skills. Because high digital skill occupations need, like those requiring research, medical treatment and architectural design occupational digital skills are more common in the services sector, more women who own digital skill who can compete to win.

High digital skill occupations more easier than men because employers usually select female to do high skill occupations easier than make. However, if those professional service female employees can not learn how to apply digital skills to do these researchs medical treatmentm architectural design professional service jobs. Then, it is also different for these professional service femal employees to raise competition in global labour professional service market. So, these professional service female employees need to learn how to apply digital to do themselves jobs in future global professional service labour market. Otherwise, if the male professional service employees can attempt to learn how to apply digital skill to do themselves jobs in order to improve efficiencies and service performance to satisfy patients, such as medical service needs, school search service needs, construction firms' building needs. Then, the owning high digital technology skillful female employees will be more easier to find the professional service jobs which need digital skill more easier than the lacking digital skill female service professionals in future global digital service professional labour market.

On the other robotic communication skill need aspect, future employers expect workers to know how to communicate with robots to work efficiently in any working environment if the employers need robotc to serve their organizations. For example, communication between the robots on factory floors, and between people and robots could allow robots to start and stopr processes based on real-time conditions around them and alert people when there is a problem, so robots could increase their own efficiency if the workers could monitor themselves and determine when they needed maintenance; efficiency would also be improved if machines and robots could make production decisions on their own by. For example, ordering new suppliers when existing inputs into a production process run low. The increase in productivity of industrial robots will likely reduce the number of manual jobs on the shop floor.

At the same time, the increased output made possible by such robots will mean that manufacturers need more people in accounting, finance, sales, advertising and other roles. The increase in putput may also drive increased employment on manufacturers' supply chains. Hence, future employers expect to employ the workers who can know how to communicate with robots to work efficiently in order to raise productivity in any working environment. It means that it the worker can know how to control and communicate with the robots to work together in the team. Then, his/her communication and controlling robotic skill will help the organization's team to work efficiently and raise productivity in order to reduce time waste and human waste and resource waste considerately. So, future shortage of communication and controlling robotic skillful workers number will increase. It has much beneficial to workers who choose to attempt to learn how to communicate and control robots to work together in any working environment team efficiently. Because future employers will like to use robots to assist manual workers to attempt to raise productive efficiency in any working environment. So, the need of employees who know how to cooperate or communicate with robots whose talent skills will be useful to any future employers.

Future global business leaders will need human machine cooperation skill. This technological skill includes artificial intelligence (AI and internet of things (IOT), will reshape our working change. These machines will participate to our daily working environment. For instance, many business leaders agree that automated systems will free-up their time as well as they also believe they'll have more job satisfaction by offloading the tasks that they don't want to do to intelligent machines.

Therefore, future leaders will expect humans and machines can work as integrated teams within their organizaton in order to their workforce and machines are already successfully working this way. So, they need to expect future employees can know or learn how to work with automated systems more easily, because many jobs will be participated by automated systems, e..g simple accounting tasks, legal administration tasks etc. clerical tasks. They will be participated with (AI) technology, it learns how to cooperate with (AI) technology to finish simplt clerical tasks efficiently.

Future workers will need have autrmated system operational skills: They include that how to operate automated systems to free -up workers' time. Workers will need to learn how to operate automated system to better with healthcare tracking devices workers will need to learn how to operate automated systems to absord and manage information in completely different ways. Workers will need to learn how to operate automated systems of smart machines to work as admin. in any orking environments. Workers need be needed to learn how to operate (AI) automated machines to mak more accurate clerical tasks or efficiencies. So, the automated system (robotic) operational skillful workers' demand and number will increase.

In the future, employers need automated machine manufacturing and service with workers cooperation reasons include that clear protocols, will need to be established if autonomous machines fail. So, they need their workers to learn how to control and manage and communicate with autonomous machines skillfully. They believe move they depend upon technology, the more they'll have to lose in the event of a cyber attack. So, skillful workers are real required to let them to know how to cooperate with autonomous machines more efficiently and easily. Computers will need to be able to decipher between good and bad commands, so future employers have much chance to need the owning automated machines operating workers to assist any robots to make more accurate good or bad decision when robots and workers have need to make immediate judgement in their any related job responsibilites aspect.

Therefore, future owning automated machines operating workers' skillful level will be high. It bases on automated machine manufacturing environment trend factor. Finally, future technology will connect the right employee to the high task at the right time. It implies that when future global employers began to accept to apply robots to help them to raise any productivities efficiently. It will influence many manufacturing positions which need to employ any proficient skillful workers who own automated machines operational skills to know how to communicate or manage or control , even supervise any robots to work in teams in any organizational manufacturing environment efficiently.

In the future, employers also expect employees to own sufficient digital vision and strategic skills, manifest among other things. They can know how to apply data to demonstrate any senior support and sponsorship digital technological skill. They expect to reduce a skill gap and avoid a lack of employee buying and a workforce culture to change in their digital technologicl manufacturing organizations. Future employers also believe outdated technology that can't work fast enough, data overload, privary and security concerns. So, it explains why it is possible that future employers also need digital working environment and automated robots machines to attempt to achieve raising productive efficient aim.

Moreover, it also explains why digital transformation need will be raised. The reasons include: They feel digital technology can gain employees' buying in , making customer experience a boardroom concern, achieving fair compensation , training and goals and strategy achievement more easily, tasking senior leaders with digital working environment change putting policies and technology to support a fully remote, flexible workforce , empowering lines of team work more efficient, teaching all employees how to code/understanding how to adopt to work with automatic machines or rots in any team efficiently. So, automate machine can raise efficiency in manufacturing society.

In conclusion, in the future business society, employees need to be stronger human machine partnerships. So , future manufacturing or service industries will have digital technology and automated machine robotic technology to assist workers to work in any working environment efficiently. They expect digital technology and automated machine robotic technology anticipation to workers' daily jobs in order to bring positive impacting to the customer experience from business owners to decision makers in marketing, customer service, research and developmnt and finance etc. They also expect technological productivity can bring positive relationship between technology and workers emerging technologies' impact on business and the way workers and automated machine work together.

Future organizational skillful
needs how to influence workforce
change to what kinds of employees

In the future whether in general organizations need what kinds of employees' skills, they expect employee individual own. It is one interesting question. The common skills that employees need to own in order to any duties to any organizational departments efficiently, e.g. human resource, marketing, administrative, logistic etc. different departments. For hospital, school, business, professional occupations etc. different organizations. Whether future school ought implement one system educational method to teach different common skills to students in order to let them to leave schools to jobs more easier.

Future employers need to create new technologies including automation and algorithms, in order to create new high quality jobs and improve the job quality and productivity of the existing work of human employees in any organizations, e.g. accounting department will need intelligence (AI) to assist account clerks to do simple repeating accounting job tasks in order to share their work load and raise performance efficiency or legal organizations will need (AI) to assist law clerks to do simple repeating legal draft or legal document revising job tasks . All future general clerical jobs will apply (AI) technological tools to assist human to job, it will produce a comprehensive platform for managing workforce change.

Hence, human manual(employees) need to learn how to adopt (AI) job participation to assist them to do different kinds of simple clerical jobs in any organizational administrative departments . They , clerical employees or white color workers need to learn how manage or dominate (AI) tool to improve job performance to be better. However, (AI) administrative workforce change, it is not only one kind of job automation change role in any physical offices. It influences future administrative clerks need change a more flexible manner, utilizing remote staffing beyond physical offices and decentralization of operations organizational workforce change.

Instead of (AI) participation to administrative job aspect, (AI) will also participate to manufacturing industry environment aspect, a new human-machine manufacturing workforce change will exist to any factories, warehouses working environment. Scientists predict that in present an average of 71% of total task hours across the industries are performed by humans, compared a 29% by machines. In this average is expected to have shifted to 58% task hours performed by humans and 42% by machines. In fact, nowadays, in terms of total working hours, no work task was yet estimated to be predominantly performed by a machine or an algorithm (AI). But, this picture is predicted to have somewhat changed with machines and algorithms (AI) on average increasing their contribution to specific tasks by 57% . For example, in the future, 62% of organization's information and data processing and information search and transmission tasks will be performed by machines compared to 46% today.

Therefore, these high technological skillful job change will bring negative influence to some demotive-skillful or low skillful labors to be dismissed, if they can not upgrade or raise or reskillgul their skill level to improve their analytical thinking , technology design and programming skills to cooperate with (AI) tools to work efficiently together in any organizational manufacturing or offie work environment. Because it will have many employers apply (AI) automation tools to participate with blue -color or whiate -color workers‘ tasks in order to raise efficiencies or improve performance in any working environment. So, it is right time to young or mid age employees need to upskill and/or reskill their rihgt type of skills to prepare future technology risch work environment changeing needs.

Future technological advances will permit an increasing number of tasks traditionally performed by humans to become automated. It seems that , such automation focused primarily on routine tasks, e.g. clerical work, bookkeeping, basic paralegal work and reporting etc. However, with the advent of big data, artificial intelligence (AI), the internet of things and ever-increasing computing power , i.e. the digital revolutions, non-routine tasks are also increasingly likely to become automated. For example, the recent development in robotics and 3D printing allow firms in advanced economies to locate production closer to domestic markets in fully aumomated factories. As a result, the future strongest incentive to automate because of their relatively higher labour costs will be reduced, when production automated will bring the negative influence to dismiss some foolish or low produtive or low skill workers , the owning high automated productive skillful workers will replace the low productive skillful workers in any factories' manufacturing environments. So, technological progress participates to raise quantity of jobs will cause result in significant job losses to low skillful workers. Because future employers will need many high automated productive employees to help them to cooperate with (AI) automated machine to work together efficiently. For example, many proportion of occupations at high risk is greatest in Germany and lowest in Korea, these countries organizations will accept to spend technology investments and education of workers to prepare future automatability manufacturing development successfully.

However, future automatability manufacturing development will bring technological unemployment in possible, due to workers need to adjust to the challenge of automation by switching tasks. Thus, preventing technological unemployment, also technological change does not just destroy jobs, but also generates new roles through its effect on productivity and the demand for new technologies. For example, it has been estimated that, for each high tech-job created in the industries , such as computing equipment or electrical machinery, some 4.9 % additional jobs are created for lawyers, taxi, drivers and waites in the local economy (Moretti, 2011). Therefore, automated will also influence service industries' job nature change, e.g. taxi drivers need to apply (AI) automated machines to assist them to drive their taxis. When the passenger tells the taxi driver where he/she wants to go. Then, the (AI automated machine will follow the GPS road direction map to be indicated how to drive the taxi to go to the destination automatically . So, future taxi driver is one assistance role to assist the (AI) automated driving tool to dominate the (AI) tool to drive the taxi to catch the passenger to arrive the destination safety in the short time in possible. For another example, future restaurant waiters will need (AI) automated machines's assistance to help them to deliver or dispatch any foods and soft drinks to send to the identified eater's table carefully in accurate and efficient service performance way from the kitchen, in especially in the busy time and many people are sitting in the large size restaurant environment. So, future, waiter roles will be the leader , they need to manage or control or supervise the (AI) robotics how to make decisions to arrange to dispatch which foods or soft drinks to the different tables in preference immediately. Also, future law clerks need to supervise or manage the law robotics how to help them to make decisions to do revision or draft or filing legal tasks in preference in order to avoid any typing words are mistaken to type on computers or revised draft in wrong way to assist manual legal clerks' mistaken words are appearanced on any legal documents. So, the law clerk future role will be the trainer role , he/ she eeds to teacher the robots how to check any words, e.g. grammers to correct them to be right grammers, or giving the accurate revision legal documents' instruction to let the legal robots to know how to revise each legal draft to prove whether which part of the legal draft will have wrong to be needed to revise.

In conclusion, future many manual workers' service or manfacturing job natures will become automated assistance to robotics. So, employees need to upgrade their skills in order to adopt new technological work nature change. So future CEO needs to prepare to learn how to apply robotic technological skills to train workers to raise efficiencies and improve performance because there are many future organizations start to apply robotic manufacturing tools to assist workers to work in any departments. Hence, future excellent CEOs must need to own AI knowledge to satisfy their organizational performance improvement need.

Skill training talent human method

Skill is an ability or effort that you need to put time to develop. Talent refers to an inborn and special ability you own it. So, when one person can attempt to learn ths kind of skill. It is possible that he can be trained to be talent person. How to Create Effective Skills Training with Career Pathing ? Your company's ability to address the skills gap is going to be the most significant issue facing HR in the next decade. Relying on the recruitment of new hires will no longer be a viable solution. Digital transformation of the workplace means that AI and automation are continually rendering skills obsolete while creating new jobs in the process. As those new roles emerge, the existing talent pool will be insufficient to meet demand. Employers will no longer be able to fall back on their default strategy of hiring new workers.

● How to improve staff skill to be talent labour ?

As the 'future of work' begins to assume a more defined shape, the majority of employers are placing more emphasis on training of their existing talent, but skills development is not moving fast enough to keep up with demand. Ongoing upskilling and reskilling can help to offset the impact on your workforce from these fundamental changes.

Creating personalized development opportunities

When it comes to developing your talent, there is no 'one size fits all' approach. To succeed in today's workplace the following steps are recommended. Your approach must become more personal, placing development at the center of your overall business strategy.

Personalized development opportunities should be offered to enhance skills acquisition. This approach enables you to provide your employees with the tools they need to acquire new skills.

As well as being targeted to the individual, employees should be able to learn in their own time. Technology can help to support this. A further, critical point to note is that learning and development is not exclusively for the C Suite but should be offered to all of your employees.

● Career pathing strategy improve employee individual skill ?

Career pathing provides a clear route into all of these options, enables you to create individual learning programs for all of your employees and offers the following benefits.

Employees create their own career paths, which are aligned with your organization's business goals. All employees are guided to understand their own strengths and weaknesses and are empowered to identify key areas for development. They are inspired to work towards vertical or lateral moves within your organization, for example, through job rotation (ie, where employees assume new tasks in a different role for a specified period before they 'rotate' back to their original post). Career pathing enables HR and management to understand and analyze employee aspirations through internal mobility programs and aligns well with your succession planning program. So, career pathing strategy is one kind of good skill to raise staff efficiency or improve performance.

Skills shortages on developing country market

Future developing countries need to develop their economy, so they need to employ many employees who own technical skills and /or soft skills. What kind of technical skills and/or soft skills , the developing countries' employees who will need to order to raise competiton in local job market ? I shall indicate the developing country China example. China is one developing country, employers will need different kinds of skillful labors to assist them to develop their businesses. However, China employers will face skillful labour shortage challenge. Although Chinese young age population is high, but many of them do not to be encouraged to learn enough skillful knowledge to fill future new skillful positions. So, the fast speed of training will be important to influence China supply and demand labour market to be more accurately as well as future China's the quality of labour demand number will be influenced to be raised after they have enough training to learn new skills.

How to solve future China skillful shortage of labour? Firstly, nowadays, China employers need to teach their employees to learn how to use and how to operate robotic skills in China's factories. AI robotic has been early developing, so they need to prepare to learn robotic management and operating technical and soft skills in order to satisfy future China factory automation industry development.

China is one world's factory for low-end products to high quality information products, high end technology and services. So, China will need many high skilled workers to assist manufacturers to manufacture many different kinds of products to export or local sale. Moreover, robotic manufacturing skillful workers will also need because robotic will be accepted to assist manual workers to work in China's any factories. This has led to greater demand for labour wirh upgraded skills and competence. So, it seems that China's orkers need to lern any high technological manufacturing knowledge, e.g. learning how to co-operate with robotics to raise productive efficiencies, which will be future many China's manufacturers' skills need intention.

So, when any one of China manufacturer invests robotics to work in its factory . Then, the China manufacturer's labours ought need to know how to co-operate with th robotics to raise productivities and efficiencies. Moreover, these China service industries, e.g. IT, software, accounting, finance, marketing and customer service management, e.g. waiter, property security, shopping center customer service etc. service occupations. In the future, robotics can also used to participate any one of these service industries‘ part of tasks in order to raise service performance. So, any one of these service industries' employees need to learn how to operate with robotics in order to achieve the most excellent servvice performances to satisfy consumers‘ needs. So, China service industries labours ought need to learn how to co-operate or manage service natural robotics to work together more efficiently because future China manufacturers will prefer to employ the labours who know how to co-operate and manage and control any service natural robotics more easily and efficiently in order to achieve the most excellent service performance to satisfy customers needs.

Hence, it seems that China manufacturing and service workers need to spend time and effort to learn how to co-operate with manufacturing natural robotics to manufacture any products in factories efficiently or deliver any cargos in warehouses more efficiently or serve customers to let them to feel excellent service performance in restaurants or shopping centers or properties or offices receiption counters. Then, when their China employers apply robotics to participate to work in factories, restaurants, shopping centers, cinemas, offices or properties reception etc. different working places . These low skillful labours will be dismissed easily, due to robotics can replace them to manufacture any products or provide services to satisfy clients' needs in order to let them to fell robotics‘ performances are more excellent to compare human service labours or their productive efficiencies are more effort to compare workers. So, future China workers need to learn how to cooperate or manage or contol with robotics to work more efficiently, if they do not expect to be dismissed easily.

In the future several occupations have been identified as the most frequent movers between all labour market states. The elementary occupations include: waiters, bar staffs, clearners, catering assistants, construction and security service workers, care workers, sales assistants and general clerks etc. So, the low educational level workers can learn these soft wkills to raise whose professional workering level to prepare to do these above positions in global elementary occupation job market.

The changes of employer were most frequent for IT programmers, doctors, electricians, carpenters, skilled workers in global labour market. These skilled occupations will have manpower shortage supply challenge, due to either people feel the educatonal level is under low. So, there has no many people have interest to know these knowledg to prepare their elementary careers. So, these kinds of low skilled occupations will have not enough human power supply to global labour job market also, the high skilled or educational job support.

Moreover, the high skilled occupations also encounter labour shortage issue. The skills in short supply related to experienced canadidates e.g. five years or more. For example, pharmaceutical , biogharma and food innovation industries. The occupational shortage roles include: Chemists, analytical scientists, product formulation, analytical development for roles in biopharma, quality control analyst includes pharmaco-vigilance, i.e. drug safety roles. The demand for engineering industry aspect which will aos increase the labour shortage includes process and design (research and development, quality control, automation, lean processes) are skillful labours need to help employers to achieve these intentions. They may include raising competitiveness, boosting productivity and skills availability. So, if future these above any one of occupation labours can not achieve these benefits to satisfy their employers' needs. Then, his/her average weekly or hourly wages will be reduced. It means that the unskilled labour under skilled labour wage can not increased more easily, even they own many year working experiences in any one of above these occupations. If the employer feels the labour is unskilled or below skilled level for any one of these occupations in these any one industry aspect, e.g. wholesale and retal , human health, education, accomodaton and food , construction, professional activities, financial service , public administration, and defence, transportation etc. occupations. Then, these industries' unskilled or below skilled level workers' salaries will be lower level to compare the higher skilled workers who work in any one of these industries.

The reason why future employers need to employ skilled labours. One explanation for slow recovery in demand in negative impact on investment is a prolonged period of high unemployment. This is led to job weekers left labour market or became unemployable due. So, future low skillful level will be one important factor to cause unemployment in society as well as nowadays labours ought need consider whether their skills are needed to improve in order to avoid future competition in job market.

2.1 Why do future labours need to learn worldwide readiness skills

Future employers need employees own worldwide readiness skills, such as reading , writing and arithmatic. Why do employees need worldwide readiness skills? In the future, high economic growth countries need high wage positions, high opportunity jobs which need a large number of skills required of job candidates of these positions " job readinss" and not " job training" , which support developments of these importance and widely desired skills won't only support the success to high-opportunity positions, but also be developed for future success in the competitive global economy. Because real-time business intelligence is needed for the talent marketplace to employ talent employees. So, it explains that it will have many future employers hope to employ owning readiness skillful employees to help them to develop their businesse intelligently. Hence, present employees ought need to hard to train readiness skills to prepare whose future employers' job requirements in the future competitive global job market.

2.2 Why these occupations need readiness skills

In the future these occupations will need to raise readiness skills. For example, mathematical science, teachers (post-secondary), management analysts, computer and information systems, managers, first-line supervisors of construction traders, solar photovoltaic installers. All of these occupations , employers need staffs to own good readiness analytic ability to help them to do more accurate real-time business intelligent decisions. The representative occupations include oral and written communication skills, project management, teamwork, marketing and creativity . Moreover, they need to own specific technology skill, deep science and math or even most business skills as well as these skills are "soft" skills more than hard skills. These kinds of occupation employees need own cooperative effort, creativity, problem solving, detail orientation and integrity personal characteristics, which are relevant across all knowledge and domains.

Therefore, in the future, science, technology,engineering and mathematics relevant occupations need to own more readniness and analytic skills more than othe kinds of occupations. Because these organizations need those professionals on knowledge acquisition, literacy analysis, synthesis and critical thinking skills that will impact their organizations to bring more critical thinking beneficial team culture. These occupational top skills will include oral and written communication skills, project management skill, team oriented skill, marketing and creativity skills, problem solving skill, detail oriented skill, self-motivated skills, management and analytical skills, coaching skill, business process modeling skills, work independent skill, strong leadership skills, management experience and business requirements gathering. All of these skills which will be future employers who need to employ these kinds employees who own these skills in preference. Also, all of these skills concentrate on soft skills more than hard skills. It seems that when above occupational applicants who own any one of thes skills, evn more than one skills. Then, he/she will have more chance to be selected to employ. Also occupation specific skills requirements are more needed to compare cross-functional skills for above of any one occupation. Because the high concentraton of cross-functional skills require " job readiness" and not " job training" for success, e.g. communicaton, integraton and presentation skills, entrepreneurialism and related skills, microsoft office software skills.

Of particular interest is communication, integration and presentation skills. These skills include ability to seek, evaluate and examine information and data create a reasoned position, present findings and make a case for or advocate for position. So, these skills are very important and they can help future applicants who expect to win any kinds of these positions easily. However, the hard skills can help these applicants to be more successful to win any kinds of these positions when they own these hard skills, e.g. microsoft offic, powerpoint, excel , word, microsoft project etc. softwares.

In conclusion, the global economy is dynamic and many of the skills required for positons in the future will need good technologies and work practices to be developed. The number of skills required t be successful in the jobs forecast to be most in demand in the future is growing. So, it explains that why future any one of these occupations which will need soft skills more than hard skills, due to organizations like to employ the employees who own managerial and analytical effort more than hard skills productive effort to assist their organizations to develop more easily.

2.3 Data -analysis skill needs

In the future, most organizations will have a number of jobs that include data analysis. Economists and labor market forecasters predict occupations need data analytical skill will need much. In addition, fast technological development means th types of technologies and applications workers in this field will need to be familiar with data analytical skill rapidly. It seems that data analytical jobs will have new job opportunity to employees with in-demand skills in future global labor market.

Why and how do employers demand for data analysis skills? Data analysis skills mean the ability to gather, analyze and draw practical conclusions from data as well as communicate data findings to others. The occupations include: data analyst, data scientist, statistician, market research analyst, financial analyst,research manager. In business career, many employers expect to employ statisticans, operations researh analysts, market research analysts and marketing specialists to assist their organizations to gather useful data from market in order to analyze and draw practical conclusions and finding the best solutions or methods to win their competitors.

Therefore, these data analysis jobs will have much need. Large size organizations with 500 or more employees were more likely than small or medium size organizations with 25 to 499 employees to plan hired data analysis positons in the future. For example, human source department will use big data to help make strategic decisions. How HR uses big data . HR will use big data for sourcing, recruitment, or selection, identifying causes of turnover and/or employee retention strategies or trends, managing talent and performance. Why organizations do not use big data. It is possible that they lack of knowledg expertise, the majority of organizatons will have data analysis positions within accounting and finance department, human resources department, business and administration department, information technology department, marketing, advertising and sales department, supply chain and operations department, research and development department, customer service department and other departments. So, future data analysis skill will need to used in different organizational departments.

However, publicly and privately owned for-profit organizations were more likely than government organizations to have data analysis positions in the marketing, advertising and sales function. Also, data analysis skills are required to different levels in any organizations , such as entry level, non-management / individual contributor level, mid-level management level, seniot management or executive level. The analyst, research analyst, market research analyst, scientist-based titles include: data scientists , research scientist, scientist, other descriptive titles include researcher, statistician, mathematician and other . So, data analysis positions will have many different skills to be selected to any one data analysis professional. For example, the data analysis professional can select either to learn the ability to interpret and communicate data analysis results skill or to learn how gathering or analyzing data skill. So, data analysis skill is not onlyone skill, it is more than one skill to let any one employee to select to learn.

Why do organizations need data analysis professionals? On workforce planning aspect, organizatons expect to let strategic direction and content of workforce needed for future business objectives easier, analyzing workforce: supply analysis, demand analsis and gap analysis more earier, developing action plan : recruiting and training plans to deal with gaps more easier, implementing action plan, monitoring, evaluating and revising plan more easier. So, organizations expect the data analysis professionsla can help them to solve these challenges, such as using of advanced technology solutons to integrate disparate planning sources; data availability and format; accessing to and understanding of the organization's data and analytics, developing business case to gain support from senior management and collaboration among HR staff, managers and executive easier. Future industries need data analysis professionals may include manufacturing health care and social assistance, scientific and technical service, finance and insurance, educational services , government agencies, retail trade, transportation and warehousing, construction, utilities, accomodation, and food services, waste management and remediation services, entetainment, and creation, real estate and rental and leasing , repair and maintenance, agriculture, forestry, fishing and hunting, personal and laundry services etc.

In conclusion, data analysis job need explains why future readiness and data analytical skills will be popular needed in globl labour market , due to these both skills are labour shortage and employers will need employees own big data readiness and data analytical both skills in order to win whose competitors more easier.

2.4 What are regional dynamic skills of global labour market demand

Businessmen expect to improve better economic environment, they will prefer to recruit the most sought after skills of intelligent employees to bring positive beneficial impact to organizations. However, technology and digisation has had a significant influence on workers. Future globalization will trend digital economic development. Hence, it will influence workers' skills to be changed also. In fact, not all changes are positive because some workers will possible lose jobs, either due to new technology replaces their jobs or they lack enough effort to improve their skills in global digital economic labour market environment.

It brings this question: What are regional dynamic skills need whn digital busines environment is growing. In fact, organizations will continue to deal with skills shortages, labour markets across the global are continually changing. so, more employers and workers will need to adopt innovate working pattern, e.g. on call jobs, freelance jobs will grow popularly. The greater flexibility afforded to employ regardly.

Finally, digitalisation includes artificial intelligence, big data , online platforms. All these new technology will influence future employees how to worker. For example, they can apply online platform to work at home conveniently. So, they do not need to go to offices. They can finish their jobs and send to their employers by email easily. This kinds of job pattern can raise efficiencies and employers do not need go to offices often.

An important implication of innovating working which needs the employees who own digital skills in order to serve organizations more efficiently. So, employers are increasingly able to access demographics that were hitherto less active in labour markets. For example, future more women are joining the labour market because part time and self employment opportunities make it easier. This kinds of job pattern can raise efficiencies and employees do not need go to offices often.

An important implication of innovating working which needs the employees who own digital skills in order to serve organizations more efficiently. So, employers are increasingly able to access demographic that were hitherto less active in labour markets. For example, future more women are joining the labour market because part time and self employment opportunities make it easier to manage family with work life. So, digital skilling needs will cause many women lose jobs in possible. If the women lack digital job skills. Because high digital skill occupations need, like those requiring research, medical treatment and architectural design occupational digital skills are more common in the services sector, more women who own digital skill who can compete to win.

High digital skill occupations more easier than men because employers usually select female to do high skill occupations easier than make. However, if those professional service female employees can not learn how to apply digital skills to do these researchs medical treatmentm architectural design professional service jobs. Then, it is also different for these professional service femal employees to raise competition in global labour professional service market. So, these professional service female employees need to learn how to apply digital to do themselves jobs in future global professional service labour market. Otherwise, if the male professional service employees can attempt to learn how to apply digital skill to do themselves jobs in order to improve efficiencies and service performance to satisfy patients, such as medical service needs, school search service needs, construction firms' building needs. Then, the owning high digital technology skillful female employees will be more easier to find the professional service jobs which need digital skill more easier than the lacking digital skill female service professionals in future global digital service professional labour market.

On the other robotic communication skill need aspect, future employers expect workers to know how to communicate with robots to work efficiently in any working environment if the employers need robotc to serve their organizations. For example, communication between the robots on factory floors, and between people and robots could allow robots to start and stopr processes based on real-time conditions around them and alert people when there is a problem, so robots could increase their own efficiency if the workers could monitor themselves and determine when they needed maintenance; efficiency would also be improved if machines and robots could make production decisions on their own by. For example, ordering new suppliers when existing inputs into a production process run low. The increase in productivity of industrial robots will likely reduce the number of manual jobs on the shop floor.

At the same time, the increased output made possible by such robots will mean that manufacturers need more people in accounting, finance, sales, advertising and other roles. The increase in putput may also drive increased employment on manufacturers' supply chains. Hence, future employers expect to employ the workers who can know how to communicate with robots to work efficiently in order to raise productivity in any working environment. It means that it the worker can know how to control and communicate with the robots to work together in the team. Then, his/her communication and controlling robotic skill will help the organization's team to work efficiently and raise productivity in order to reduce time waste and human waste and resource waste considerately. So, future shortage of communication and controlling robotic skillful workers number will increase. It has much beneficial to workers who choose to attempt to learn how to communicate and control robots to work together in any working environment team efficiently. Because future employers will like to use robots to assist manual workers to attempt to raise productive efficiency in any working environment. So, the need of employees who know how to cooperate or communicate with robots whose talent skills will be useful to any future employers.

Future global business leaders will need human machine cooperation skill. This technological skill includes artificial intelligence (AI and internet of things (IOT), will reshape our working change. These machines will participate to our daily working environment. For instance, many business leaders agree that automated systems will free-up their time as well as they also believe they'll have more job satisfaction by offloading the tasks that they don't want to do to intelligent machines.

Therefore, future leaders will expect humans and machines can work as integrated teams within their organizaton in order to their workforce and machines are already successfully working this way. So, they need to expect future employees can know or learn how to work with automated systems more easily, because many jobs will be participated by automated systems, e..g simple accounting tasks, legal administration tasks etc. clerical tasks. They will be participated with (AI) technology, it learns how to cooperate with (AI) technology to finish simplt clerical tasks efficiently.

Future workers will need have autrmated system operational skills: They include that how to operate automated systems to free -up workers' time. Workers will need to learn how to operate automated system to better with healthcare tracking devices workers will need to learn how to operate automated systems to absord and manage information in completely different ways. Workers will need to learn how to operate automated systems of smart machines to work as admin. in any orking environments. Workers need be needed to learn how to operate (AI) automated machines to mak more accurate clerical tasks or efficiencies. So, the automated system (robotic) operational skillful workers' demand and number will increase.

In the future, employers need automated machine manufacturing and service with workers cooperation reasons include that clear protocols, will need to be established if autonomous machines fail. So, they need their workers to learn how to control and manage and communicate with autonomous machines skillfully. They believe move they depend upon technology, the more they'll have to lose in the event of a cyber attack. So, skillful workers are real required to let them to know how to cooperate with autonomous machines more efficiently and easily. Computers will need to be able to decipher between good and bad commands, so future employers have much chance to need the owning automated machines operating workers to assist any robots to make more accurate good or bad decision when robots and workers have need to make immediate judgement in their any related job responsibilites aspect.

Therefore, future owning automated machines operating workers' skillful level will be high. It bases on automated machine manufacturing environment trend factor. Finally, future technology will connect the right employee to the high task at the right time. It implies that when future global employers began to accept to apply robots to help them to raise any productivities efficiently. It will influence many manufacturing positions which need to employ any proficient skillful workers who own automated machines operational skills to know how to communicate or manage or control , even supervise any robots to work in teams in any organizational manufacturing environment efficiently.

In the future, employers also expect employees to own sufficient digital vision and strategic skills, manifest among other things. They can know how to apply data to demonstrate any senior support and sponsorship digital technological skill. They expect to reduce a skill gap and avoid a lack of employee buying and a workforce culture to change in their digital technologicl manufacturing organizations. Future employers also believe outdated technology that can't work fast enough, data overload, privary and security concerns. So, it explains why it is possible that future employers also need digital working environment and automated robots machines to attempt to achieve raising productive efficient aim.

Moreover, it also explains why digital transformation need will be raised. The reasons include: They feel digital technology can gain employees' buying in , making customer experience a boardroom concern, achieving fair compensation , training and goals and strategy achievement more easily, tasking senior leaders with digital working environment change putting policies and technology to support a fully remote, flexible workforce , empowering lines of team work more efficient, teaching all employees how to code/understanding how to adopt to work with automatic machines or rots in any team efficiently. So, automate machine can raise efficiency in manufacturing society.

In conclusion, in the future business society, employees need to be stronger human machine partnerships. So , future manufacturing or service industries will have digital technology and automated machine robotic technology to assist workers to work in any working environment efficiently. They expect digital technology and automated machine robotic technology anticipation to workers' daily jobs in order to bring positive impacting to the customer experience from business owners to decision makers in marketing, customer service, research and developmnt and finance etc. They also expect technological productivity can bring positive relationship between technology and workers emerging technologies' impact on business and the way workers and automated machine work together.

In the future whether in general organizations need what kinds of employees' skills, they expect employee individual own. It is one interesting question. The common skills that employees need to own in order to any duties to any organizational departments efficiently, e.g. human resource, marketing, administrative, logistic etc. different departments. For hospital, school, business, professional occupations etc. different organizations. Whether future school ought implement one system educational method to teach different common skills to students in order to let them to leave schools to jobs more easier.

Future employers need to create new technologies including automation and algorithms, in order to create new high quality jobs and improve the job quality and productivity of the existing work of human employees in any organizations, e.g. accounting department will need intelligence (AI) to assist account clerks to do simple repeating accounting job tasks in order to share their work load and raise performance efficiency or legal organizations will need (AI) to assist law clerks to do simple repeating legal draft or legal document revising job tasks . All future general clerical jobs will apply (AI) technological tools to assist human to job, it will produce a comprehensive platform for managing workforce change.

Hence, human manual(employees) need to learn how to adopt (AI) job participation to assist them to do different kinds of simple clerical jobs in any organizational administrative departments . They , clerical employees or white color workers need to learn how manage or dominate (AI) tool to improve job performance to be better. However, (AI) administrative workforce change, it is not only one kind of job automation change role in any physical offices. It influences future administrative clerks need change a more flexible manner, utilizing remote staffing beyond physical offices and decentralization of operations organizational workforce change.

Instead of (AI) participation to administrative job aspect, (AI) will also participate to manufacturing industry environment aspect, a new human-machine manufacturing workforce change will exist to any factories, warehouses working environment. Scientists predict that in present an average of 71% of total task hours across the industries are performed by humans, compared a 29% by machines. In this average is expected to have shifted to 58% task hours performed by humans and 42% by machines. In fact, nowadays, in terms of total working hours, no work task was yet estimated to be predominantly performed by a machine or an algorithm (AI). But, this picture is predicted to have somewhat changed with machines and algorithms (AI) on average increasing their contribution to specific tasks by 57% . For example, in the future, 62% of organization's information and data processing and information search and transmission tasks will be performed by machines compared to 46% today.

Therefore, these high technological skillful job change will bring negative influence to some demotive-skillful or low skillful labors to be dismissed, if they can not upgrade or raise or reskillgul their skill level to improve their analytical thinking , technology design and programming skills to cooperate with (AI) tools to work efficiently together in any organizational manufacturing or offie work environment. Because it will have many employers apply (AI) automation tools to participate with blue -color or whiate -color workers' tasks in order to raise efficiencies or improve performance in any working environment. So, it is right time to young or mid age employees need to upskill and/or reskill their rihgt type of skills to prepare future technology risch work environment changeing needs.

Future technological advances will permit an increasing number of tasks traditionally performed by humans to become automated. It seems that , such automation focused primarily on routine tasks, e.g. clerical work, bookkeeping, basic paralegal work and reporting etc. However, with the advent of big data, artificial intelligence (AI), the internet of things and ever-increasing computing power , i.e. the digital revolutions, non-routine tasks are also increasingly likely to become automated. For example, the recent development in robotics and 3D printing allow firms in advanced economies to locate production closer to domestic markets in fully aumomated factories. As a result, the future strongest incentive to automate because of their relatively higher labour costs will be reduced, when production automated will bring the negative influence to dismiss some foolish or low produtive or low skill workers , the owning high automated productive skillful workers will replace the low productive skillful workers in any factories‘ manufacturing environments. So, technological progress participates to raise quantity of jobs will cause result in significant job losses to low skillful workers. Because future employers will need many high automated productive employees to help them to cooperate with (AI) automated machine to work together efficiently. For example, many proportion of occupations at high risk is greatest in Germany and lowest in Korea, these countries organizations will accept to spend technology investments and education of workers to prepare future automatability manufacturing development successfully.

However, future automatability manufacturing development will bring technological unemployment in possible, due to workers need to adjust to the challenge of automation by switching tasks. Thus, preventing technological unemployment, also technological change does not just destroy jobs, but also generates new roles through its effect on productivity and the demand for new technologies. For example, it has been estimated that, for each high tech-job created in the industries , such as computing equipment or electrical machinery, some 4.9 % additional jobs are created for lawyers, taxi, drivers and waites in the local economy (Moretti, 2011).

Therefore, automated will also influence service industries' job nature change, e.g. taxi drivers need to apply (AI) automated machines to assist them to drive their taxis. When the passenger tells the taxi driver where he/she wants to go. Then, the (AI automated machine will follow the GPS road direction map to be indicated how to drive the taxi to go to the destination automatically . So, future taxi driver is one assistance role to assist the (AI) automated driving tool to dominate the (AI) tool to drive the taxi to catch the passenger to arrive the destination safety in the short time in possible. For another example, future restaurant waiters will need (AI) automated machines's assistance to help them to deliver or dispatch any foods and soft drinks to send to the identified eater's table carefully in accurate and efficient service performance way from the kitchen, in especially in the busy time and many people are sitting in the large size restaurant environment. So, future, waiter roles will be the leader , they need to manage or control or supervise the (AI) robotics how to make decisions to arrange to dispatch which foods or soft drinks to the different tables in preference immediately. Also, future law clerks need to supervise or manage the law robotics how to help them to make decisions to do revision or draft or filing legal tasks in preference in order to avoid any typing words are mistaken to type on computers or revised draft in wrong way to assist manual legal clerks' mistaken words are appearanced on any legal documents. So, the law clerk future role will be the trainer role , he/ she eeds to teacher the robots how to check any words, e.g. grammers to correct them to be right grammers, or giving the accurate revision legal documents' instruction to let the legal robots to know how to revise each legal draft to prove whether which part of the legal draft will have wrong to be needed to revise.

In conclusion, future many manual workers' service or manfacturing job natures will become automated assistance to robotics. So, employees need to upgrade their skills in order to adopt new technological work nature change.Such as Amazon is one super ecommerce organization, it ought need all different departments staffs prepare to ungrade their different kinds of skills, e.g. reading skills for KDP publish review book tasks in order to let authors can have more attractive topics to let readers can buy any Amazon ebooks to read. So, reading skills may be the essential element to Amazon any one book review staff in itself online publish business. Even ebook ecommerce platform design whether Amazon KDP publish online platform can attract to let global readers to feel enjoyment to read its ebooks , it is one essential element to help Amazon to increase e-readers number every day, so Amazon publish platform needs to improve itself ebook reading platform design improve to attract many e-readers like to choose Amazon e-reading platform to read Amazon any authors ebooks. So, E-reading platform design staffs need to upgrade themselves e-reading platform design skills. So, Amazon knolwedge workers number must need to increase in order to continue to improve its e-platform design to attract global e-reading customers or e-shoppers choose to buy Amazon any kinds of products.

Reference

Moretti, E. (2011) local labor market in O, Ashentelter and D. Card (eds.) handbook of labor economics, Elsevier, North Halland.

CHAPTER THREE

Technology how helps businesses changes labor effort

Although, (AI) technology will be popular to applied to different jobs, but it still needs social acceptance to replace some human jobs. Today, it is increasingly common for people to use robots in various situations at home and in retail stores, hotels and hospitals. Robots are classified into several types based on their functionality (service and utility robots or those designed to communicate with humans) and appearance (humanoid robots or mechanical robots). The types of robot to which every country attaches particular important in the advance of robotics, reflects the sense of values and preferences of its population . Thus, (AI) will be applied to replace human to do these above different kinds of job nature. For example, U.S. has the highest level of robot utilization at home and an retail stores with its people being the most enthusiastic about the future use of robots. Otherwise, Germany shows a strong tendency to consider robots for industrial purposes, and its people feel strong to the presence of robots in their households. Japanese accepts to apply" human aid robot" that can communicate with humans and they have a high level of familiarity with robots.

Hence, it implied those three countries have accept (AI) to replace human to do any these kinds of job duty and it will influence these three countries‘ workers lose their old occupations and who will unemployed absolutely, due to many (AI) robots replace them to do their job duties in the future. Also, US will have many retail service workers or retail warehouse workers are unemployed. Germany will have many manufacturing industry's workers are unemployed. Japanese will have many communication industry workers are unemployed, such as telephone service, shopping center services etc. different kind of service industry's service staffs . It will cause these kind of workers' competitive abilities are lost in themselves countries' jobs that require such skills include software developers, court judges, nurses, high school teachers, dentists and university lecturers, these occupations are still difficult to be replaced by (AI) robots.

Are robots taking our jobs or making them? In fact, our societies will have unemployment challenges, even (AI) technology has not created before. However, after (AI) robots invention, some of human jobs will be replaced and it can raise many low skillful and low knowledge level worker unemployment number. However, I think that high productivity driven by increasingly powerful IT -enabled machines is the causes of global labor market problems and accelerating technological change will only make those problems worse.

IT technology brings this question: Are robots killing human's jobs or benefiting human's jobs? I suppose that there is a limited amount of labor to be done. The implication is that technology can create unemployment by displacing workers, such as (AI) invention, because the more efficiently worker work (using machines or (AI) robots), the loss work there is for workers to do. Even, any new jobs will be better done by machines or (AI) robots, and unemployment will still skyrocket. How do we know that humans will always be better at some work, or more importantly, enough work, than machines or (AI) robots, e.g. human drivers drive more safe or careful to compare (AI) robot drivers. But, the challenge is that it is not ensure that (AI) robots drivers must not drive careless to cause the chance of accident occurrences more than human drivers. However, technological change can be beneficial to innovation, automation and increasing productivity for businesses.

Consequently , it may seem machines can hurt wages and job for low skillful, less educated workers. Also, high educated workers are likely as less educated workers to find themselves displaced and devalued, and more education may create as many problems as it solves. Thus, in negative influence, automation effects on particular jobs shift workers to other jobs that are equally or more desirable. Workers may be highly compensated for possessing human capital that is specialized to a labor market. If technological advance is very rapid, such as (AI) invention, causing a large and very rapid drop in demand in a large labor market, the economy may not be able to absorb the sudden surplus of labor in a short period of timer when (AI) robots are popular to replace some workers to do some occupations in global societies.

For example, self-driving vehicles threaten to send truck drivers to the unemployment office. Computer programs can now write journalistic accounts of sporting events and stock price movement. There are even computers that can grade essay revolutionize some part of teaching jobs. Hence, (AI) robots will have possible to replace human brain to do any judgement, argument, and mind job duties. It implies some occupations which need human' mind will be threaten by (AI) robots, e.g. author, accountant, nurse, engineer. Thus, (AI) robots will have possible to replace some professional and high educated workers' jobs in the future.

But, technology can create new nature of jobs in possible. For example, a 60 minutes program indicated technology is putting new categories of jobs in the sites (sic) of automation, the 60% of the workforce that makes its living gathering and analyzing information. Also, recession: technology kills middle -class jobs that overall technology is eliminating for more jobs than it is creating by (AI) technology. Hence, human's brain work may be assisted by 60% of (AI) gathering and analyzing information for some occupation , e.g. space scientists, ocean scientists, earth scientists etc.

However, I believe the (AI) invention and human job competition may influence global productivity change. Productivity is economic output per unit of input, the unit of on input can be labor hours(labor productivity), but if (AI) robots replace human job, then the unit of input may be (AI) machine hours (AI) robot productivity or all production factors including labors, machines and energy (total factor of productivity). Producing more output with less input can take several forms.

The traditional notion of productivity is a form reorganizing production and/or using better or more technology to produce more output per worker hour. But when (AI) robots invention, the form can be reorganizing production and/or using better or more (AI) robots to produce more output per (AI) robot hour. Hence, if the firm apply (AI) robots to produce its products. Then , productivity improvements in the firm may result in less workers employment, due to (AI) robots replace more worker number to achieve more productivity improvement, it has economic benefits (less factor of production) , but more production in long term.

Thus, (AI) robots can help any firm to achieve productivity improvement in long term, for example, if unproductve farmers move to the city and start working for high-tech. manufacturers. The shift effect can be more dynamic and disruptive as low-productivity industries lose out in the marketplace to high -productivity industries and the compositional mix of the economy changes. Thus, in the long term (AI) robots can also be beneficial to high productivity industries to bring the mix of economy positive changes.

Moreover, automation will also produce some new jobs in firms that sell the new robot or other labor-saving technology. This means that, in general, there will be shift in the economy in the direction of higher-skill and higher wage jobs. Even if the (AI) robot invention country, US becomes a leader in (AI) robots producing productivity-enhancing technology, it will experience a growth in jobs serving foreign (AI) robots product buyers. Hence, (AI) robots can also create (AI) salespeople, (AI) manufacturing workers , (AI) inventors, scientists, (AI) software designer etc. occupations, when if all society does is move workers from insurance firms, restaurants and car factories to robot factories, productivity will have remained the same to create job needs for insurance, restaurant and car manufacturing worker service occupations for (AI) software designer, (AI) service robots manufacturer, (AI) service robot seller etc. related (AI) service robot product occupation created in (AI) robot technology job market. Hence, (AI) invention also create new (AI) technology job chance. (AI) impacts management job market.

In future, organization management will be changed from (AI) introduction. Division of labor will change and collaboration among humans and machines will increase. Companies will have to adapt their training, performance and talent acquisition strategies to account for a new found emphasis on work that hinges on human
judgement and skills, including experimentation and colloboration.

How (AI) impacts any organizational administrative management work? (AI) 's greatest impact will be on administrative coordination and control tasks, such as scheduling, resource allocation and reporting, (AI)-driven will place a higher premium on what we call " judgement work", the application of human experience and expertise to critical business decisions and practices when information available is insufficient to suggest a successful course of action. This kind of work will require new skills and mindsets; replacing people with machines is not goal in itself. When, artificial intelligence enables cost-cutting automation of routine work, it also empowers value -adding augmentation of human capabilities.

Thus, administrative and routine tasks, such as scheduling, allocation of resources, and reporting, will within intelligent machines, responsibilities that have long been reserved for humans. For instance, a typical store manager or a lead nurse at a nursing home must constantly juggle shift schedules, accounting for staff members' absense owing to illness, vaction, time or sudden departures. Many of these tasks will be automated by (AI). Imagine (AI) writing management monthly reports, it is not a distant dream. Leading news providers and Wall street banks are now using (AI) report generators to write news and analytical reports by drawing on quantitative data. The associated press, for example, expanded its quarterly earnings reporting from approximately 300 companies to nearly 3,000 with the help of (AI) powered software robots, freeing up journalists to conduct more investigative and interpretive reporting. For another example, Jobalime, a job-placement site, uses intelligent voile analysis algorithms to evaluate job applicants. The algorithm assesses paralinguistic elements of speech, such as tone and inflection, products which emotions a specific voice will elicit, and identifies the type of work at which an applicant will likely excel. In the future , (AI) machines can be applied to assist some kind of office administrative jobs duties. It's attractive to office managers to achieve more accurate judgment to do any administrative matters when who can be assisted from (AI) machines. Thus, managers need to spend time to learn how to apply (AI) machine to assist them to do more accurate judgement, and better informed choices. (AI) robots can be applied to improve the speed quality and cost of available products and services, instead of applying on productivity improvement and administrative improvement aspects. Thus, they may also displace large numbers of workers. This, possibility challenges the traditional benefits model of trying health care and retirement savings to jobs.

In an economy that employs dramatically fewer workers to deliver benefits to displaced workers. For example, the worldwide number of industrial robots has increased rapidly over the past few years. The fall prices of robots, which can operate all day without interruption, make them cost- competitive with human workers. In special consideration, in the service sector, computer algorithums can execute stock trades in a fraction of a second, much faster than any human. As those technologies become cheaper, more capable, and more widespread, they will find even more applicants in an economy.

Consequently, (AI) technology brings unemployed number increasing many businesses continued automating their operations rather than hiring additional workers. A trend among technology companies that receive massive valuations with relatively few workers. For example, in 2014 year Google was valued at $370 billion with only 55,000 employees, a tenth the size of AT & T's workforce in the 1960 year. Hence, if automation technologies like robots and artificial intelligence make jobs less secure in the future, there needs to be a way to deliver benefits outside of employment " flexi security" or flexible security is one idea for providing healthcare, education and housing assistance whether or not someone is formally employed.

In conclusion, (AI) and robots technology will raise unemployment to some occupations when (AI) replaces same industries' workers job duties in our societies in the future, but it also create new jobs to raise employment in any related (AI) robots and automated machine products in (AI) manufacturing. (AI) design, (AI) sale self-related industry, when (AI) replaces same industries' workers' job duties.

(AI) journalism, media publishing, digital communication technology trend

How to apply (AI) technology in digital communication journalism media, publishing industry? Some scientists indicate future (AI) and digital technology may consist such as: voice driven assistants, emerge. For example, Amazon e book publish applying digital technology and (AI) auto printing technology to sell e books to let readers to listen any e book content by (AI) voice driven speaker when they turn on computer to read e book contents; capable phones start to unlock the possibilities of 3D image of mobile story telling. New smart wearables include ear buds that handle instant translation and glasses that talk and hear. China and India will become a key focus for digital growth with innovations around payment online identity, and artificial intelligence. Thus, future (AI) technology can be applied to 3D image mobile story telling, online payment method to dealt online transaction publishing industry.

Thus, future (AI) technology can be applied to online e book publishing industry to make sound books to let readers feel more attractive . Such as Amazon publish has published sound e books to attract readers to choose to read any its books from online. Also, (AI) technology can also be applied to communication industry. For example, some online pure-play news, opinion and entertainment websites. It is a digital communication media, e.g. online journalism blog (AI) technology can be applied to visual storytellers to let online book readers to enjoy to listen to watch and send any online electronic book contents more attractive. Thus, future (AI) technology will be popular to assist any electronic book publishers to publish visual and sound talking storybook to let readers who can watch motive image and listen and read words from e books more attractive.

Thus, (AI) technology can be applied to internet ecommerce publishing or media industry to help any electronic book publishers to publish sound, image motion electronic book to attract global readers to read, even (AI) technology can be applied to digital entertainment industry, e.g. electronic 3D image virtual video games, computer games. It can be also applied to education industry, e.g. the first true digital native generation and are the native speakers of the digital language of computers to let student to learn different languages or translate words to compare to classroom learning more easily. It can be also applied to communication industry, e.g. (AI) mobile phone. Hence, it seems (AI) technology can be applied to publishing, communication, education , entertainment etc. different industries in the future. (AI) technology will be one kind of tool to satisfy human's daily life needs in the future and these industries has one characteristics is that they need to apply internet to operate to operate to do online business.

Thus, it has three trends of (AI) technology and internet technology need to be linked to achieve one kind of attractive technological business to satisfy client's needs. These three trends as below: All consumer trends involve the internet. It will be many consumer's online habits, shopping, working, socializing, watching TV, studying, travelling, listening. Thus, (AI) music, eating and exercising are just a few examples. This is happening because human usually use mobile broadband or Wi-Fi, rather than cables. Thus, (AI) technology will be applied to mobile to satisfy client's need absolutely.

● How (AI) influences organizational change

Consequently creative and social intelligence will be in even greater demand as (AI) makes in management and the workforce. This development will represent a long term trend in labor markets , one characterized by intensifying demand and reward for social skills with a growing desire for creative capabilities, managers will seek to fashion of ideas and hypotheses from inside and outside of the enterprise to shape solutions to their most pressing business problems. Thus, (AI) will influence overall organizational team members who have chance to participate any decision to make more accurate business judgment.

Many managers mistakenly view judgment work as only an individual discipline, failing to appreciate that it can also involve decide interpersonal and organizational practices. In more complex settings, judgment is typically a collective outcome of individuals' and teams' diverse perspectives, insights and experiences. And often , the resulting choices are better informed than decisions that an individual would have arrived at on his or her own.

Thus, when any organizations apply (AI) technology to assist managers to gather data and ideas to make any judgment. In these cases, organizations can create the conditions for effective collective judgment by establishing structures , such as " shadow advisory boards" that prompt managers and employees to source and synthesize multiple perspectives. Thus, a traditional organization (firm) might freshen its thinking is t put together a shadow advisory board, comprised of young, digital people who can apply (AI) machine assistance to make judgment work more accurate whether related to people development, problem-solving or strategizing and innovating for considerable degrees of creative and social intelligence.

Thus, on the one hand, (AI) technology machine augmentation and automation can give these advantages to human (organization managers) , e.g. developing people and community, solving problems and collaborating, coordinating and controlling work, shaping strategy and leading innovation. Besides, on the other hand, the next generation managers need have these individual attitude to treat intelligent machines to be as colleagues.

When, judgment is a human skill, intelligent machines can accelerate human learning that supports it, assisting in data -driven simulations, scenarios and search and discovery activities. Focuses on judgment work, some decisions require insight beyond what data can tell them. This is the sweet sport for human judgment, the application of experience and expertise to critical business decisions and practices. Thus, managers will also need to find ways to learn how to use digital (AI) technologies to tap into the knowledge and judgment of partners, customer external stakeholders and role models in other industries after the (AI) machine had been implemented to the organization.

Future works change:Automation, employment and productivity

2.1 How (AI) influences employment

Human future " micro to macro" industry trends will be affected business strategy and public policy by (AI) technology. In the future (AI) technology will influence those six themes: productivity and growth, natural resources, labor markets, the evolution of global financial markets, the economic impact of technology and innovation and urbanization. However, (AI) technology will bring economic benefits of tackling gender inequality, a new global competition, Chinese innovation and digital globalization.

Nowadays, advances in robotics artificial intelligence, and machine learning are in a new age of automation, as machines match or outperform human performance in a development to any countries. For example, automation of activities can enable businesses to improve performance by reducing errors and improving quality and speed, and in some cases achieving outcomes that go beyond human capabilities. For example, some research indicated automation could raise productivity growth globally by 0.8 to 1.4 % annually; more than 2,000 work activities across 800 occupations. When less than 5% of all occupations can be automated using demonstrated technologies about 60% of all occupations have at least 30% of constituent activities that could be automated. Many occupations will change that will be automated away: Activities most susceptible to automation involve physical activities, in highly structured and predictable environments, as well as the collection and processing of data. They are most prevalent in manufacturing , accommodation and food service and retail trade and include some middle-skill jobs. For example, such as natural language processing is a key factor. Beyond technical feasibility, the cost of technology competition with labor including skills and supply and demand dynamics, performance benefits including and beyond labor cost savings, and social and regulatory acceptance will be affected by (AI) automation technology. Thus, (AI) automation will impact to influence global employment in those aspects as below:

Firstly, assuming that people are displaced by automation will find other employment. The anticipated shift in the activities in the labor force is of a similar order as the long-term shift away from agriculture and decreases in manufacturing share of employment. Both of manufacturing and agriculture industries which would be accompanied by the creation of new types of work not foreseen at the time.

Secondly, for business, the performance benefits of automation are relatively clear. Thus, the businessmen have opportunities for their micro economies to benefits from the productivity growth potential and macro economies to benefit to encourage continued progress and innovation , investment and market incentives. At the same time, employers must innovate policies to help workers and institutions adapt to the impact on employment.

This will likely include rethinking education and training, income support and safety nets , as well as support for those dislocated, when employees need to leave themselves homes to move to other cities to learn new (AI) automation works. Thus, individuals in the workplace will need to engage move comprehensively with machines as part of their everyday activities, and acquire new skills that will be in demand in the new automation age. Consequently , the scale of shifts in the labor force over many decades that automation technologies can be a similar order to the long -term technology -enables shifts in the developed countries' workforces away from agriculture in the 21 th century. Those shifts did not result in long-term mass unemployment because they were accompanied by the creation of new types of work not foreseen at the time. However, human will still be needed in the workforce when the total productivity gains are caused by (AI) technology.

2.2 What occupations will be influenced by (AI) technology.

In the future, scientists predict that these occupations will be influenced by (AI) technology mostly. They include : retail salespeople, food and beverage service workers, language or translation teachers, health practitioners. Since these work activities have a more relevant occupations are made up of a range of activities with different potential for (AI) automation . For example, a retail salesperson will spend more time interacting with customers, stocking shelves , or ringing up sales. Each of these activities is distinct and requires different capabilities to perform successfully.

Thus, these job activities have similar simple control characteristics. Simple activities include greet customers, answer questions about products and services, clean and maintain work areas, demonstrate product feature process sales and transactions. All these activities can have similar simple activities in order to (AI) machines can be learn how to do these activities from (AI) technology . For example, the capability perception includes sensory perception, cognitive capabilities, such as retrieving automation, recognizing known patterns(supervised learning), logical reasoning problem solving.

Thus, (AI) machine is such human, which has feeling and emotion, such as social and emotional sensing, judgement reasoning methods, natural language understanding and physical capabilities, such as mobility , navigation, gross motor skill, fine motor skills. It seems that the future, (AI) human invents machines which will have these human characteristics to do human similar behavioral job duties more easily and efficiently. It implies these above human occupations will be replaced by (AI) human invention machines in the future. Due to (AI) creation, it is possible to cause unemployment number of these above workers will increase because (AI) machines can do their similar job behavioral activities.

Consequently, employers won't need to employ many of these skillful labor. Otherwise, they can buy less number (AI) machines to attempt to do whose job activities more easily and efficiently. So, it seems (AI) machines will have more high work performance to replace these occupation workers' work performance. Finally, these occupation worker unemployment number will only increase when the (AI) machines had been invented to achieve to do their work behavioral activities absolutely success in the future.

2.3 Whether (A) technology machine labor will replace human worker more or assist human worker more

There is no single agreed definition of a robot how outcome of a task that is completed without human intervention. When some definitions require the task to be completed by a physical machine moves and respond to

its environment, other definitions use the term robot in connection with tasks completed by software , without physical embodiment.

However, to answer the question : Whether (AI) technology machine labor will replace human worker more or assist human worker more. I shall indicate some examples to let readers to judge whether (AI) technology can create new jobs or reduce old jobs.

Firstly, I shall explain what (AI) function is. (AI) is a service robot that performs useful tasks for humans or equipment excluding industrial automation application . Thus, the classification of a robot into industrial robot or service robot is done according to its intended application. It is also a personal service robot or a service robot for personal used for a non commercial task, usually by lay persons . Examples are domestic servant robot, and pet exercising robot. It is also a professional service robot or a service robot for professional used for a commercial task, usually operated by a properly trained operator. Examples, are cleaning robot for public places, delivery robot in offices or hospitals, fire-fighting robot, rehabilitation robot and surgery robot in hospitals. Thus, these functions will be future (AI) application to our daily life necessaries or business necessaries.

However, some authors agree (AI) will bring negative outcomes of automation, due to raise competiveness, reduce human job nature. Otherwise, other authors argue (AI) will bring positive outcomes of automation, due to raise productivities, job creation, assist humans work.

On the positive outcome hand, robots can increase productivity . This is particularly important for small-to medium sized businesses both are in developed and developing countries economies. It also enables large companies to increase their competitiveness through faster product development and delivery. Increased use of robot is also enabling companies in high cost countries to re shore, or bring back to their domestic base parts of the supply chain that will have previously outsourced to sources of cheaper labor. Currently , the greater threat to employment is not a automation, but an inability to remain competitive. Automation has led overall to an increase in labor demand and positive impact on wages. The reason is that the middle-income/middle-skilled jobs have reduced as a proportion of overall contribution to employment and earnings leading to fears of increasing income inequality, the skills range within the middle income bracket is large. Thus, robots are driving an increase in demand for workers at the higher -skilled and with a positive impact on wages. This issue is how to enable middle-income earners in the lower-income range to unskilled or retain. Finally, the (AI) positive impact supporter who argue the future will be robots and humans can work together.

However, on the negative outcome hand, robots can substitute labor activities, but don't replace jobs. They believe that less than 10% of jobs are fully automatable. Increasingly , robots are used to complement and augment labor activities, the net impact on jobs and the quality of work is positive. Automation can provide the opportunity for humans to focus on higher-skilled, higher-quality and higher-paid tasks. Robots can improve productivity when they are applied to tasks that which perform more efficiently and to a higher and more consistent level of quality than humans. For example, increased productivity is enabling some firms, such as Whirlpool, Caterpillar and Ford Motors company in the US restructure their supply chains, bringing back parts of the manufacturing process to the country of origin. Thus, productivity gains due to robotics and automation are important not just at the company level, but also for build industry and nation competitiveness.

I suppose that productivity can be raised. What are the impacts of robots on employment? Firstly, the main focus of development has been on personal entertainment, which does not drive worker productivity (manufacturing production). When the internet (information and communication technology (ICT)) innovation. This is borne and by findings that manufacturing productivity, which has been driven by innovations in automation rather than consumer technologies, has government strongly than productivity in the services sectors of the economy in most nature economies. It seems (AI) automation will create many jobs in internet communication entertainment game industry. For example, many young people like to use internet to play any electronic games from computer or mobile at home or outside home conveniently. Thus, (AI) automation will increase demand to be invented to any new entertainment game from internet channel. It will need to employ many (AI) entertainment game inventors to create many automation entertainment games. Thus, (AI) automation in internet entertainment game industry will need human (AI) entertainment game inventors to invent the knowledge-based capital of (AI) automation entertainment games. The (AI) entertainment game inventors will need own research and development skills, form specific skills, organizational know-how skills, databased knowledge, design and various forms of intellectual property to do these (AI) automation entertainment game invention occupations in the future.

International Federation Of Robotics(2016) indicated that China will be as a major robotics manufacturer and user of robots, benefiting from jobs created by robot manufacturing and productivity gains from robot use. Chins had sold of robots to any one single market every year since 2017 year. The Chinese government has included a focus on robotics in its 10 year strategy. In order to achieve its target of a robot density of 150 units per 10, 000 workers by 2020 year. Thus, Chinese companies will have to install around 650,000 new industrial robots between 2016 to 2020 year, 2.5 times more than installed globally in 2015 year.

Hence, China (AI) manufacturing industry will need to employ many workers . It implies (AI) manufacturing industry will create many new occupations in China. Also, ministry of economy, trade and industry (2015) also showed that Japan currently has the largest stock of industrial robots in operations, primarily in the automation industry. Driven by a rapidly aging population and low productivity rates, the Japanese government has sights on a 20-fold increase in the use of robots in the non-manufacturing sector and a three-fold growth rate of labor productivity in the service sector both by 2020 year. Thus, it also implies Japan will need many robots to be provide to service industry. Due to robots will provide to serve any businessmen's clients. Thus, it is possible that the service workers won't be dismissed as well as it is depended on the serving job nature to decide whether Japan's service workers can still serve to their employer when the service (AI) robots are applied to whose employers.

Consequently, it seems that (AI) can create employment, Ministry of economy, trade and industry (2015) showed that such as China will develop the major (AI) automation manufacturing industry. The (AI) employers will need to employ many workers to manufacture any these different kinds of (AI) robots to satisfy China or overseas individual or business buyers needs. But, (AI) can also cause unemployment to the low skillful service workers. Such as if Japan some service businesses choose to buy any (AI) service robots to replace their service staffs to serve their clients. It is possible that the service staffs will be dismissed, due to (AI) robots can do such as their same service job duties to achieve better service performance. Thus, today, it is increasingly common for people to use robots in various situations at home and in retail stores, hotels and hospitals these service industries. Robots are classified into server types based on their functionality (service and utility robots or those designed to communicate with humans) and appearance (humanoid robots or mechanical robots). The type of robot, to which each country allocated particular importance in the advance of robotics, reflects the sense of values and preferences of its population. Thus, if the country has high population needs to use robots, then they will influence either more new jobs creation or more old job loss in the country's (AI) manufacturing or (AI) service industries both. For example, Japan respondents often associate the term " robot " with humanoid robots that can communicate with human and they have a high level of familiarity with robot. The US has the highest level of robot utilization at home and in retail stores with its people being the most enthusiastic about the future use of robots. Germany shows a strong tendency to consider robots for industrial purposes and its people feel strong effort to the presence of robots in their households.

In conclusion, to judge whether how (AI) will influence the country's employment to be better or worse. It will depend on the country home buyers (users) or business buyers (users) how to use (AI) for their daily needs. If the country , such as US retail stores need to use (AI) , it will have possible to reduce some or many retail service workers. Even, if the country , such as Japan has many home users need to use (AI) , it will not influence the employment market. Otherwise, it will raise (AI) salespeople numbers. Even, if the country, such as Germany and China will have many (AI) manufacturers, then it will create many (AI) manufacturing occupations for these (AI) manufactory workers.

Consequently, (AI) robots manufacturing and service needs will have positive or negative impact to any country's employment. It will depend on the (AI) service provision and service workers' job nature as well as the manufacturing workers of (AI) knowledge level to decide their employment chance in their country's employment market.

- Can (AI) impact human leisure need changes?

Human need concern this question: Will artificial intelligence (AI) reduce some human jobs in order to instead of replacing machines to do? Due to artificial intelligence is the ability of machines to do thing, that people would require intelligence. For example, artificial intelligence machine man driving(self-driver), it (AI) machine man driving research is an attempt to discover and describe aspects of human intelligence that can be simulated by driving machine functions. Alternatively, (AI) mathematical research may be another viewed as an attempt to develop a mathematical theory function to describe the abilities and actions of things (natural or man-made) exhibiting intelligent behavior and server as a design of intelligent calculation machine function.

Why do humans need artificial intelligence machines to instead of traditional human service job? For example, can artificial intelligence machine man (self-driving) driver drive to replace human driver? I shall compare the differences between humans and computers : The characteristics of humans are good at recognizing various things, either seen before or not, recognizing the relationship patterns between things.

Human thinking is common sense reasoning, combining all types of sensory input, acting appropriately in novel situations, learning new things and changing behavior patterns, making decisions , even when given incomplete information, working with noisy, incomplete information gathering behaviors . However, characteristics of computers are good at: The tasks humans do naturally are extremely difficult for a computer program as intelligent, which must be able to do the same kind of tack as humans do naturally.
Hence, (AI) is an combination of many different success and technologies: Linguistics - computational and socio, philosophy-logic, philosophy of mind and of language, electronical engineering -image and speech processing, pattern recognition, robotics, machine learning, neural networks, optimization scheduling, management information system and decision making. So, it is possible that (AI) can impact human job nature to instead of human working behavior in the future.

How can (AI) influence labor market?
● How can human society job nature
to be changed to artificial intelligent society?

From the first intelligent perspective reason view point, artificial intelligence is making machines " intelligent" acting as humans expect people to act. Artificial intelligence has ability to distinguish computer responses from human responses, it owns knowledge to solve expert problem. From another research perspective reason view point, artificial intelligence is the study of how to make computers do things which, at the moment, people do better (Rich & Knight, 1991, p.3).
(AI) researchers are native in a variety of domains, e.g. formal tasks (mathematics, games), tasks (perception, robotics, natural language, common sense reasoning), expert tasks (financial analysis, medical diagnostics, engineering, scientific analysis and other areas).
From the second business perspective reason view point, (AI) is a set of many powerful tools, and methodologies for using those tools to solve business problems. From a programming perspective reason view point, (AI) includes the study of symbolic programming problem solving and search .
From the third human technological perspective reason view point, today's computer can do many well-defined tasks, for example, arithmetic operations, are much faster and more accurate than human beings.

However, the computers' interaction with their environment is not very sophisticated yet. How can human test whether a computer has reached the general intelligence level of a human being? Can a computer convince a human interrogator that it is a human? But before thinking of such advanced kinds of machines, human will start developing our own extremely simple " intelligent" machines.

So, it is possible that human society job nature will to be changed to artificial intelligent society when (AI) technology is developed to the mature stage in the future.

- Why does human need artificial intelligence machines?

One of major division in (AI) is between humans who think (AI) is the only serious way of finding out how we (human) work and human who want companies to do very smart things, independently of how we (human) work. This is the important distinction between cognitive scientists vs engineers. One of another major division in (AI) is between symbolic (AI), which represents information through symbols and their relationships. Specific Algorithms are used to process these symbols to solve problems or deduce new knowledge and connectionist. So (AI) , which represents information in network. Biological processes underlying learning, task performance and problem solving are imitated from human mind behaviors. Thus, it is possible that artificial intelligence machines can do the better judgicious behavior to compare human.

- How does artificial intelligence influence future working changing in automation employment and productivity aspects?

In the automation changing influence aspect, as companies increasingly use robots on production lines or algorithms to optimize their logistics manage inventory, any carry out other core business functions. Technological advances are creating a new automation age in which ever-smarter and more flexible machines will be deployed on an ever larger scale in the marketplace. However, researching artificial intelligence with how influences human working nature. We need to answer these questions: How will automation transform the workplace? What will the implications for employment? And what is likely to be its impact both on productivity in the global economy and on employment?

Advances in robotics, artificial intelligence, and machine learning are growing in a new age of automation as machines match or outperform human performance in a range of work activities, including ones requiring cognitive capabilities. What factors are determined the changing in workplace adoption by artificial intelligence innovation? What advantages are automation? Automation of activities can be enabled businesses to improve performance by reducing errors and improving quality and speed, and achieving outcomes that go beyond human capabilities.

Some scientists indicated based on their scenario modeling. They estimated automation could raise producing growth globally by 0.8 to 1.4 percent annually. Almost, the activities people are paid almost $16 trillion in wages to do in global economy have the potential to be automated by adopting currently demonstrated technology. According to their analysis of more than 2,000 work activities across 800 occupations. When less than 5% of all occupations have of least 30% of activities that could be automated. They also indicated that technical economic and social factors will determine automation. Continued technical progress, for example, in areas such as natural language processing is a key factor beyond technical feasibility , the cost of technology, competition with labor including skills, and supply and demand dynamics, performance benefits including and beyond labor cost savings and social and regulatory acceptance will affect (alter) the scope of automation.

Other some scientists also indicate U.S. country for example, the anticipate shift in the activities in labor force of a similar order of magnitude as the long term sight away from agriculture and decreases in manufacturing. Share of employment in the United States both which were achieved. So, those factors can influence why artificial intelligence technology needs. So, it is possible that future agriculture and manufacturing both industries will apply (AI) technology manufacturer-kind of job nature to raise productivity instead of farmers, fruit picking workers, farming transportation labours as well as factory manufacturing workers and supervisors etc. human-kind of job nature.

- Is artificial intelligence possible to replace labor ?

Not just intelligence, but also debating, if machines are capable of having a conscious minds. Artificial intelligence has those characteristics as below:

On functionalism aspect, artificial intelligence inputs mental states, sensory inputs, (beliefs, desires being in pain feeling) and behavioral outputs. Since mental states are identified by a functional role, which are thoughts to be manifested in various systems. Even, perhaps computers which are physical devices with electronic substrate that inform computations on inputs to give outputs similar to brains which are artificial intelligence composed of part any intrinsic relationship to each other. Thus, artificial intelligence activities is not the whole itself, but into parts or on external influence on the parts.

On dualism aspect, artificial intelligence is a set of views about the relationship between mind are matter. On materialism aspect, it builds the only thing that exists is matter, including consciousness.

On biological naturalism aspect, it is similar a human brain than feels pains makes mental situation. So, artificial intelligence is similar biologist which might to be excited to human labor work. Hence, it seems artificial intelligence can change (alter) or replace human labor work of nature in possible in the future.

- Can (AI) technology replace human labour nature of work?

On technological innovation reason view point, the history development of artificial intelligence studying the intelligence is one of most ancient scientific discipline. The history development of artificial intelligence what aims to achieve human use to sense, learn remember and think, logic probability, decision making and calculation develop from mathematics, instead of replacement human labor functions.

Artificial intelligence history development aim is the scientific analysis of skills in connection and practice with the appearance of computers from 1950 year beginning. The artificial intelligence (AI) can deal with the ultimate challenges. How can (either biological or electronic) mind sense, understand and manipulate a world that is much simple and more complex than itself? And what if would human like to construct something with such capabilities?

The general-purpose software of the early period of (AI) were only able to solve simple tasks effectively and failed when which should be used in a wider range or an more difficult tasks. One of the sources of difficulty was that early software had very few or mix knowledge about the problems which handled, and activities successes by simply syntactic manipulation. Moreover, the other difficulty was that many problems that were tried to solve by the (AI) were untreatable.

The early (AI) software whether trying step sequences based on the basic facts about the problem that should be solved, experimented with different combinations till which found a solution. From the end the 1960 year, developing the so-called expert systems were emphasized. These systems had (sue-based) knowledge base about the field which handled. Till to the beginning of the 1970 year, (Prolog) the logical programming language was born, which was built in the computation realization of a version of the resolution calculus. (Prolog) is a remarkably prevalent tool in developing expert systems (on medical, judiciary and other scopes), but natural language parsers were implemented in this language. Then, in 1981 s, the Japanese announced the fifth generation computer system project, a 10 years plan to build an intelligent computer system that use the (Prolog) language as a machine code. Nowadays, (AI) can be applied any industries, such as car manufacturing industry can use (AI) technological machine-men manufacture car, instead of replacing human labors in factory. Even, in the future, using (AI) machine-men drivers can drive any private cars or public transportation tools, instead of replacing human drivers, e.g. bus, train, tram, ferry etc. Also in the future, machine-men can replace housewives to serve families to do housekeeping clean job , e.g. cleaning toilets, bathrooms, kitchens, even cooking functions at home. So (AI) machine-man can reduce housewives works at home. Moreover, (AI) machine man can take care old people , when who are living at homes or elder care centers.

So, it seems artificial intelligence (AI) will be possible developed to manufacture a new generation machine-man to assist (serve) families to do any simply cleaning or cooking jobs at homes. Moreover, the overall demand of (AI) general social needs will also rise, such as security, driving transportation tools, restaurant cleaning, elder centers care service etc. So, it seems that individual or families or social needs of (AI) will be increase in the future. Thus, it will influence macro economy growth (GDP) if there are large house family consumer group and hotel or bus or taxis or ferry etc. different business consumer group demand any artificial intelligence machine numbers increasing. Then, the artificial intelligence products and material manufacturers must need to buy many artificaial intelligence materials to produce any kinds of artificial intelligence machines to prepare to satisfy consumer individual needs. Consequently, macro economy will grow to the owned artificial intelligence development countries, e.g. US, China, UK.

First, On machine-man satisfactory demand aspect view point, it makes computers that think, it is the automation of activities. We associate with human thinking: like decision making, learning. It is the act of creating machine that perform function that require intelligence when performed by people. It is the study of mental faculties through the use of computational models. It is the study of computations that make it possible to perceive, reason and act. It is a branch of computer science that is concerned with the automation of intelligent behavior. It is anything in computing service that human don't yet know how to do property.

Second, on thought aspect artificial intelligence means systems thank think like humans, systems that think rationally.

Third, on behavioral aspect, artificial intelligence systems that act like human and that systems act rationally. However, the basic objective of (AI) is to represent human's thought processes in computation . These machines are supposed to exhibit behavior that. It is performed by a human being, would be considered intelligent. However, some authors feel (AI) has disadvantages, such as it is not creative, it is excited in the use of sensory devices, it can't make use of a very wide context of experiences and it does not use common sense.

For speech recognition and understanding function needs example, (AI) can be applied in speech recognition and understanding function, which (AI) speech or voice recognition is a data input method. For example, the computer recognizes and understands one (or a few) word commands. Speech understanding on the other hand is the computer's ability to understanding a spoken language. That is , the computer understands the meaning of sentences, an paragraphs through (AI).

So, (AI) can be attempted to learn human language how to speak. It is similar to translate human language skill, instead of actual human speaking skill. Also, (AI) can assist handicap learning or language student how to listen different languages by machine-man sounds from computers more accurately. So, it seems that it (AI) can replace human language teachers speaking function and can change teaching language nature of job in language speaking and listening education industry.

Nowadays, new technology development is popular. However, artificial intelligence is one kind of new technology choice among different technologies innovation. So it brings this question: Is artificial intelligence technology value to invest? To answer this question. I shall indicate some other new technology developments to compare (AI) technology development to judge which has urgent needs to achieve human expectation nowadays.

For example, why is green peace interested in new technologies? New technologies features prominently in our ongoing campaigns against genetic modified crops and number power. However, which are also an integral part of our solutions to environmental challenges, including renewable energy technologies, such as solar, wind and wave (water) power energy as well as waste treatment technologies, such as mechanical, biological treatment.

It seems humans need concern how to apply (AI) technology to solve environment pollution challenges in our future. So, environment protective, agriculture, natural energy technology will be popular demand to attempt to apply (AI) technology to solve their challenges or apply (AI) to assist to develop their industry.

Advances in artificial intelligence (AI) technology and related fields have opened up new markets and new opportunities progress in critical areas, such as health, education, energy, economic development, social welfare and the environment pollution.

(AI) automation will continue to create wealth and expand the global economy development in the future. However, when many will benefits that growth won't be costless and will be accompanied by changes in the skills, that workers need to increase productivity in the economy and structural changes in the economy. So, in the skills that workers need to succeed in the economy and structural changes.

I shall indicate why aggressive policy action will be needed to help Americans who are disadvantaged by these changes , due to (AI) technology is caused. For automation industry change example, artificial intelligence (AI) capabilities will enable automation of some tasks that have long required human labor. These artificial intelligence technology introduction can increase new opportunities for individuals. The economy and society, but (AI) has also the potential to disrupt be current livelihoods of many Americans. However, (AI) leads to unemployment and increase in inequality over the long run depends not only on the (AI) technology itself, but also on the institutions and policies that are changed.

Thus, it is possible that (AI) technology will raise some countries unemployment number if the employer apply (AI) technology workers to work instead of human labor in their factories, but it can also raise productivities for these employers.

- Can (AI) influence global economy growth?

Technological progress is main driver of growth of GDP per capita, allowing output to increase faster than labor and capital . However, technology can increase productivity, but also decrease the number of labor hours needed to create a unit of output. So (AI) causes unequal to labor wage decreases, even reduces the number of labor to manufacture, e.g. artificial intelligence technology of automation car manufacturing industry; clothing manufacturing industry; plane manufacturing etc. high technology of artificial intelligence manufacturing method. But (AI) should be potential environment benefit, although it raises unemployment ratio. Moreover, it can rise production , due to many skilled craft were replaced by the combination of machines and lower-skilled labor. The result of (AI) technology introduction , it causes output per hour risen when inequality declined, driving up average living standards, but the labor of some high-skill workers was no longer as valuable in the market. Otherwise, if (AI) technology is continue developed to be success. Some routine intensive occupations will be loss, which focused on predictable, e.g. easily programmable tasks, such as switchboard operators, filing clerks, travel agents, and assembly line workers would be particularly replaced by new (AI) technology. However, at the same time, (AI) technology development will bring these benefits: improvement in education (training (AI) technology scientists) , due to (AI) manufacturing technology needs are raising to businesses and institutional changes, such as the reduction in unionization and raising in the minimum wage to the (AI) manufacturing technology skilled labor in factories.

Because (AI) technology is not a single technology, but rather a collection of technologies that are applied to specific tasks, the effects of (AI) will be felt unevenly though the economy. It will bring some tasks will be most easily automated than others , and some jobs will be affected more than others, both negatively and positively. Finally, new jobs are likely to be directly created in areas , such as the development and supervision of (AI) as well as indirectly created in a range areas though out the economy as higher incomes lead to expanded demand.

However, if (AI) technology could dominate global labor markets. If labor productivity increases, do not influence into wage increases, then the large economic gains brought about by (AI) technology could be increased wealth inequality, due to employers can reduce production cost, but workers (labors) wages will not be increased, even will be decreased. Hence, it seems the (AI) technology will bring disadvantages to labor market to cause unemployment or reduce wages in possible, although it can reduce employer individual salary (wage) expenditure and it can raise productivity.

- How can artificial intelligence impact global economy growth?

Artificial intelligence (AI) technology is a branch of computer science that aims to create intelligent machines that work and react like humans. So, (AI) is a technology that appears to impact (influence) human preference by learning, understanding complex contents, enhancing humans in executing both routine and non-routine tasks. In the future, (AI) technology that can be virtual personal assistant, as well as it may exist, such as robots with human-like processing capabilities.

How can (AI) technology impact global economy growth over the next 10 years? During this time period, (AI) technology is predicted to have wide-ranging applications including: Machine learning that automates analytical model building by using algorithms that allow machines to operate without human assistance.

In global education aspect, potential applications include predicting cause-and-effect relationships from biological data, identifying new drugs, self-driving cars, and protecting against fraud, improved natural language processing that allows computers to continue to better analysis, understand and generate language to interface with humans using natural human languages. For example, transcribing notes dictated by physicians, automatically drafting articles and translating text and speech. So (AI) technology can be applied to education aspect to improve humans' knowledge level.

In visual art aspect, (AI) machine vision that allows computers to identify objects, scenes and activities in images. Current applications of (AI) machine vision include providing objective descriptions for the blind seeing(visual) needs.

We except the economic effects of (AI) technology to include both direct GDP growth from sectors that develop or manufacture. (AI) technology and indirect GDP growth through increased productivity in existing sectors that employ some form of (AI). If (AI) technology is an increasingly critical component of more products, it will become an integral part of many people's lives. Thus, (AI)'s ability to influence economic activity, rather than the economic or development status of the region. (AI) has the potential to impact income classes and to bring significant gains to both developed and developing countries. For example, (AI) has the potential to optimize good production around the world by analyzing agricultural regions and identifying what is necessary to improve crop yields.

In estimating the future economic effects by (AI) technology innovation, it is important to note that it is challenging to accurately predict which applications of (AI) will ultimately be commercially successful. In micro level economic influence, we need to apply methodologies to estimate the economic effects of investment in firms developing (AI) technology since investment levels in a technology are a telling sign of the future potential of that (AI) technology.

How (AI)'s development may affect the global economy over the next ten years. In fact, (AI) technology has the potential to affect business across the global in a wide range of industries in ways only a number of technologies have done in the parts. For example, (AI) technology's expected to be a useful tool for enhancing human capabilities and in some instances replacing functions, such as driving a car, adoption of broadband internet, mobile telephone, industrial robotic automation have served to enhance human capabilities.

However, significant public debate has focused on projections of (AI) technology's effect on the labor force. However, large companies prefer to invest in (AI) technological industry. For example, face book's (AI) research lab., google machine intelligence lab. and micro soft machine learning and artificial intelligence research division are all making advances in (AI) technology and investing in the industry's top talent. Additionally, between 2010 year and 2015 year, nearly $5 billion in venture capital funding invested in firms across the global developing and employing (AI) technology (Facebook (AI) Research).

- How can artificial intelligence impact on workplace?

Modern information technologies and the labor economy growth of machines is powered by artificial intelligence have already strongly influenced the world of work in the 21 ST century. Computers, algorithms and software simplify every tasks and it is impossible to image how most of our life could be managed without them. How can be the information economy characterized by exponential growth replaces the most production industry based on economy of scales? What will the future world of work look like and how long will it take to get? Will the future world of work be a world where humans spend less time earning their livelihood? Alternatively, are mass unemployment, mass poverty and social distortions also possible scenario for the future, where robots, artificial intelligence systems play an increasingly central role? These questions concern how artificial intelligence further development . Can influence labor economy growth on workplace ? When the labor market has widespread impact on intelligence property, information technology, product liability, competition and labor and employment laws.

How (AI) technology impacts on labor workplace.

The future influence any organizations how labor economies use of (AI) can be analyzed, such as deep machine learning is based on a set of model high level data. Unlike human workers, the machines are connected the whole time in workplace. If one machine makes a mistake, all autonomous systems will keep this in mind and will avoid the same mistake the next time.

Over the long run intelligent machines will win against every human expert. Production robots have been replacing employees because of the (AI) technology. They work more precisely than humans and cost loss. Creative solutions like 3D printers and the self learning ability of these production robots will replace human workers, the automatic data recording and data processing, traditional back office activities are no longer in demand. Autonomous software will collect necessary information and will send it to the employee who needs it. Additionally, dematerialization leads to the phenomenon that traditional physical products are becoming software. For example, CD or DVDs are being replaced by streaming services. The replacement of traditional event ticket, e-travel ticket service products or hard cash will be the next step, due to the possibility of payment by smartphone. So, (AI) technology will impact human's daily life consumption behaviors in the future. For another example, transportation tools, such as boats and ferries and private vehicles will use sensors and navigating without human input. Taxi and truck drivers will become obsolete, the stock store applies to stock managers and postal carriers of the delivery is distributed by (AI) machine delivery method.

Nowadays , (AI) is a technology almost as old as the computer industry itself, it is similar with the advent of personal assistants function to businesses and personal promotion channel, such as (Amazon's Alexa, Apple's Siri, Google's Assistant) image recognition (face book), personalized recommendations (Netflix , Amazon). Those innovations have been driven by a increase in processing power, lower cost hardware, and the exploding creation and availability of data. It seems, (AI) technology can impact global customer service management method.

How to forecast economic impact modeling to (AI) will affect global economy? Can human forecast business revenue growth and job creation (or destruction) based on (AI) applied to customer relationship management (CRM) activities? In addition to the economic impact on (AI) or (CRM) which can include an estimate of the economic impact attributable to sales forces customer base. What can economic benefits be brought to (CRM) from (AI) technology?

Artificial intelligence(AI) comprises a set of technologies that use natural language processing, machine learning, knowledge graphs, and other tools to answer questions, discover insights and provide recommendations. Computer systems can use (AI) hypothesize and formulate possible answers based on available evidence can be trained through the ingestion of vast amounts of content, and automatically adapt and learn from (AI) self mistakes and failures.

So, any business organizations (customer service departments) can provide efficient and effective customer relationship management of excellent customer service quality if which applied (AI) technology system. The different type of (AI) systems include: (AI) system platforms, machine learning (AI) based data preparation and enrichment tools, machine vision/ image recognition, voice speech recognition, text analysis and natural language processing, bots , e.g. face book website and virtual digital assistance solutions, social media pattern analysis , sentiment analysis, advanced numerical analysis (e.g. IOT streaming , machine logs), supporting technologies, knowledge base dialog management, Q&A processing etc. different (AI) technology system customer relationship management (CRM) tools.

(AI) (CRM) of activity can include these categories, such as: corporate marketing, marketing operation, field marketing, customer support, digital commerce, customer analytics, customer influenced product or service design, product or service pricing, finance information, presentation, customer billing, inventory , logistics and fulfilment support, partner management etc. different CRM tools.

(AI) technology of CRM has been carrying on plan different stages to achieve CRM personal assistant tool for businesses. The stages are such as, in the beginning stage of (AI) projects in place, implement now, pilot phase next year in the final stage of (AI) customer relationship management tools are foreseeable future. So, this CRM technology has been improved to plan in different stages every year to prepare to achieve full capacity of CRM service quality for businesses to use in the future.

Hence, how to develop an estimate prediction of the economic impact (AI) technologies could have CRM activities, which depends on gathering macroeconomic information on business revenue and the basic marketing of business revenue and the basic markup of business expenses by major functions (customer support, marketing and sales , production etc.)

An economic impact model that can gather data together and forecast the results how (AI) artificial intelligence technology brings (CRM) customer relationship management benefits to businesses, e.g. surveys investigation includes IT spending by sample countries, GDP and population estimates and forecasts, revenue per employee and ratios of IT spend to GDP. Surveys (questionnaire questions) of forecast results are influenced by (AI) impact can include: results are projected from surveys and rely on estimates are made by respondents on the expected financial improvements in categories of (AI) –assisted customer relationship management activities. The forecast assumes that these estimates are correct; financial estimates are based on estimates of "first year" improvement from full (AI) implementation; forecasts are from planning to implement any artificial intelligence of customer relationship management (CRM) projects, the improvement forecast is of categories of activity , e.g. corporate marketing , digital commerce, and customer analytics. They are not estimates of ROI for the (AI) software. They rely on conservative estimates to which each of these entities might affect company revenue, expenses or productivity. They also rely on estimates of the penetration of software in customer relationship management activities . Net new jobs created are based on the ratio of new revenue to jobs required to support that revenue . They can assume that 50% of the net new revenue will support increases in labor and the rest will go for capital and other operating expenses that may replace jobs lost to automation.

In the future, some of the ways in micro economic benefits to any organizations. (AI) technology is expected to impact CRM activities include: Spending up sales cycles, improving lead generation and qualification solving customer support problems faster (raising service quality), helping companies improve brand campaigns and recognition, lowering costs of support calls when increasing resolution rates, lowering the cost of recruiting employees and partners, increasing revenue from optimized product marketing, optimizing price, distribution logistics and preventing loss through fraud detection. So, micro economic benefits view point, it seems that (AI) CRM technology can raise any companies economic benefits for care term.

Artificial intelligence enables machines or the in-build software to behave like human beings which allows these decisions and act. The advent of (AI) is leading , talking, making decisions and act. The advent of (AI) is leading to new technologies advances and transforming the economic and employment opportunities for humans in a positive way. (AI) related technologies can facilitate our live. For example, industrial robotics, robotic medical assistants, smart games, financial forecasting software, big data analysis, algorithms in health and bioinformatics, pilotless cargo places, drone ambulances and general purpose and workplace robots and others. (Disruptors technologies: Advances that will transform life, business and the global economy).

Artificial intelligence also known as computational intelligence is defined as " the human –like intelligence exhibited by machines or software. It is theorized that intelligence of humans can be described and intelligence machines or software can simulate it. These machines software can be reasonable , learn, perceive and process information, like human mind and thus facilitate human life. They can think and act for us. So, artificial intelligence is an interdisciplinary field of study including computer science, neuroscience, psychology, linguistics and philosophy.

However, (AI) research and developments have economically impacted many industries, such as robotics, telecommunications, computer applications , health, finance, heavy manufacturing, transportation, aviation, e-service and e-commerce, military , music and movie, toys and games entertainment etc. industries.

In fact, many ideas, systems and technologies have been developing in the world of (AI) technology. However, which are net called or considered (AI) products, rather which are mentioned with their specific names, such as smart graphics, machine learning, e-commerce etc. (i.e. this is called (AI) effect).

Nowadays, (AI) related industrial applications will replace most human power in fields, including call centers, customer services and air cargo transportation. (AI) technologies also help weather forecasting based on repeated rainfall pattern (data) recognition, through robotics (i.e. floor cleaning, moving lawns etc.) transporting people and products with unmanned vehicles, sending space unmanned smart shuttles, developing robotic arms, predicting market values in stock exchanges by internet, making homes safer, helping elderly and disabled using robotic servants etc.

Among the (AI) related technologies , there are a few that significance for the impact on society and especially on digital economy . (AI) is particularly influential in machine learning. Such as robotics, transportation, finance, health and bioinformatics, e-commerce , e-games, big online data gathering and internet-of-things. For example, machine elearning is based in bioinformatics and robots that can learn new skills for better caregiving in healthcare. What is machine e-learning? Machines can e-learn from e-data gathering, coming up generalizations and making decisions to act in certain ways from internet.

There are important applications , such as e-machine perception, electronic online natural language learning processing, online search engines, online bioinformatics, online brain –computer interface, online game playing, online robot locomotion, online advertising, online computations finances, online health monitoring, online DNA classification and decision making, online in chemistry –cheminformatics . So, online machine learning can positively impact productivity and it can enhance information and analytical system from (AI) online channel.

What is robotics? Robotics is one of the most strongly influenced fields in (AI). For example, heavy manufacturing industries, robots and used and man power is replaced for effectiveness, precision, and accuracy, especially in respective or dangerous tasks, including welding, assembling , picking and placing .

So, robots can acquire new skills or adapt the changing dynamic environment. Also, artificial intelligence can be applied in developing transportation. For example, automated vehicles, driver assistance systems , safety systems, collision avoidance systems and public transportation. Moreover, (AI) technology has proven to produce some of the best tools to predict stock market fluctuations from internet data gathering method. It's predictions are based on ever-evolving predictions algorithms and systems learn new models and make connections between historical data and new data to measure stock market trading more accurate from internet data gathering channel.

In health field, especially in health data processing , analysis, decision making support and medical diagnosis. So, online data can show which patients will need what treatment and what alternative drugs could be used more accurate from (AI) online data gathering method. Bioinformatics is an interdisciplinary field combining statistics, (AI) online technology can help in discovering data patterns and modeling through the application of machine learning, artificial neural networks and genetic algorithms. For example, further (AI) technology development of human genome project of online data sequences.

Online shopping can be facilitated by virtual assistants developed through (AI) technology and these assistants can offer the best advice. (AI) online purchase coming after every product image recommendations and personalization bring important revenue to shopping online sites, like Amazon . Smart computer graphics and games, artificial intelligence is useful in smarter computer, graphics, scene modeling , scene rendering processes in order to create, for example, effective human –robot interactions , online machine learning, online strategic games techniques etc. online computer related (AI) software.

So, online big data analysis and big data does have a critical need in the world of online intelligence machines and software in our future. In other words, (AI) offers online technology to enable online big data analysis to provide industrial organizations with valuable information for effective decision making in short time. For example, what IBM's Watson achieved: this machine used 200 million of structured and unstructured content with a special technology of hypothesis generation, massive evidence gathering, analysis and scoring from internet channel.

Finally, (AI) online technology another related internet invention (internet of things) (IOT) is the network of machines or objects connected through internet. These connected objects can sense their internal and external environment, communicate with each other, can send critical data and finally can make decisions to act or correct their environment from (AI) online technology. For example, factories can monitor and automatically change production processes, hospitals can monitor and regulate the health conditions of their patients , schools can collect data from facilities and cars can send data to car makers from (AI) online technology.

Partner predicts that (IOT) market will create about trillion amount value by 2020 year. Although machines collect big data from their environment, whether which gain an insight or learn from these online data largely depends on the (AI) online machine learning principals and (AI) online technology. In 2013, Mckinsey estimated that disruptive technologies closely related with potential economic impact in 2025 year between $7.1 to $13.1 trillion amount (automation of knowledge work, advanced robotics, autonomous or near-autonomous vehicles).

Can (AI) technology influence China economy? Could China workers be affected and jobs made up of routine work activities and predictable? Will programmable tasks be particularly impact to China employment market ? When impact on labor market is likely to be gradual at the aggregate level, it can be sudden and dramatic at the level of specific work activities, rending some job obsolete fairly. Overall (AI) technology will raise digital skills when reducing demand for medium incomer inequality for China workers. It seems (AI) technology's effect on productivity could be crucial to China's future economic growth as the population ages are increasing.

In China, some biggest technological companies driving significant investments in research and development. Moreover, China is one of the leading global (AI) technology development county. However, China will need to focus on building its innovation capacity. For example, United States and United Kingdom are currently producing more influential (AI) technological research. However, if China planed to achieve (AI) technology success, it's traditional industries will need to develop technical know-how –to and overcoming implementation costs prepare to develop (AI) . When (AI) technology is introduced into China society, China government needs to raise concerning ethical, legal, technological security etc. business questions. Also, surrounding issues include privacy, discrimination, legal liability and regulation. It aims to encourage overseas investors to choose to invest (AI) technological industry to raise GDP growth and manufacturing industries income growth for long term in China. If China encouraged overseas (AI) technology investment in its country. It is possible to influence China employment market to be changed. Because (AI) technology will impact to influence China people daily life. Due to (AI) technology is introduced to China society, many rich people will prefer to spend to buy any high (AI) technological products for entertainment or learning or machine man driving etc. daily necessity activities. Then it will raise GDP growth and will raise (AI) manufacturers or related-(AI) technological manufacturers profit. It is beneficial to China because it can become one high knowledgeable and (AI) technological economical society. But it will bring bad influences to raise unemployment chance for the low skillful labor. In labor economy aspect influence , how (AI) technology can influence China low skillful labor unemployment ratio raising. The raising low skill labor unemployment reason is because China low skillful human labors are argued or are replaced by (AI) technology creating new challenges to introduce to influence China society of simply human manufacturing job nature to be changed to be high (AI) technology manufacturing job nature in any China factories. Moreover, when (AI) technology introduction to China, it will cause other related social challenges in China. The varied (AI) related challenges, including the difficulty of creating safe and reliable hardware for sensing and affecting (transportation and education), the challenges of gaining public trust, a low resource comities and public safety and security, the challenges of overcoming fears or marginalizing humans in China employment and workplace and the risk of diminishing interpersonal trust because the low skillful labors won't believe any China employers will give chance to employ them , due to (AI) technology will replace their skills and man manufacturing of productivity is much less to compare to (AI) technology manufacturing method.

Are future nature of jobs changed to computerization from (AI) technology? Where are the probability of computing occupations from (AI) technology influence? What is expected impacts of future computing on labor market from (AI) technology influence? John Maynard Keynes's frequently cited prediction of widespread technological unemployment " du to our discovery of means of economic the use of labor outrunning the pace of which we can find new used of labor" (Keynes, 1933, p.3).

In the future, (AI) technology will impact some nature of occupations to change computing. This chance will also influence some countries' economic change. For example, some factory human labors hand routine manufacturing tasks will be changed to computerization of routine manufacturing tasks by (AI) technological machine men hand manufacturing method. it will cause a structured shift in the labor market, with workers reallocating their labor supply from middle-income manufacturing to low-income service occupations.

Arguably, this is because the manual tasks of service occupations are less computerization, as who require a higher degree of flexibility and physical adaptability. So, (AI) technology will influence the human hand labor skillful occupation nature of task cheaper , such as vehicle manufacturing , ship manufacturing, computer manufacturing, steel manufacturing, television, radio etc. home electronic products of heavy machine industry change. Due to (AI) technology machine man will be proper to be used to manufacturing these electronic products when the (AI) technology innovation can develop to the mature stage. Then, any countries manufacturers will choose to use (AI) technology machine man, instead of human hand production.

Supposing the future prices of computing are fallen, seriously, problem solving skills are becoming relatively productive, explaining the substantial employment growth in manufacturing occupations, involving cognitive tasks where skilled labor has a comparative advantage, as well as the increase education needs for (AI) technology computing of machine man subject study.

Prediction of education needs for (AI) technology student numbers will increase, due to manufacturing industry needs many (AI) technology students in future employment market. Another (AI) technology influence if the future (AI) technological innovation, e.g. machine man manufacturing or machine man service industries will both increase demand, then with more sophistic software technologies will be disrupted labor markets by marketing workers redundant.

For publishing industry, what is striking about the case in paper book publishing industry will be unpopular? Due to the electronic book publishing industry will be popular, e.g. Amazon publish . (AI) technology can influence paper book manufacturing method which is replaced by machine man electronic book manufacturing method as well as it will cause the computerization is no longer confined to routine manufacturing tasks. Due to (AI) machine man manufacturing technology will be proper to be used to manufacture any products in short time efficiently and effectively , e.g. electronic book products. In the future, if it is fact to occur this case, such as (AI) technological machine man manufacturing method will be adopted (applied) to manufacture electronic books or any products in possible. (AI) technology will cause many manufacturing workers are unemployed. It is beneficial to employers, who can reduce to spend much wages expenditure to employ manufacturing workers, but it will cause many manufacturing workers loss jobs and reduce income to support whose families lives. It will cause social challenges, e.g. increasing stealing crimes if the manufacturing workers had not other skills to find other jobs to do easily. So, manufacturers need to concern over technological unemployment which will be hardly future phenomenon if who decided to dismiss all manufacturing workers, due to (AI) technology machine men replace to them.

If (AI) technology can be innovated to produce any kinds of machine man to serve any service or manufacturing industries successfully. Then, it will bring these questions: Can future that workers be influenced to be automation employment and productivity by (AI) technology influence? Does it impact to influence the (AI) technology countries' productivity and growth and natural resources development and labor markets and evolution of global financial markets and economic impact of technology and innovation and urbanization etc. issues? How will automation transform the workplace? What will be the implication for employment? What is likely to be its impact both on productivity in the global economy and on employment?

In fact, automatic of activities can enable businesses to improve performance by reducing errors chance and improving quality and speed, and same cases achieving outcomes that go beyond human capabilities. Some economists indicate (AI) technology would give a needed boost to economic growth and prosperity have of the working age population in many countries. Based on the scenario modeling, they estimate automation could raise productivity growth globally by 0.8 to 1.4 % annually. They also indicated that almost half the activities people are almost $1.6 trillion in wages to do in the global economy have the potential to be automated adapting current demonstrates technology, according to their analysis of more than 2,000 work activities across 800 occupations. When less than 5% of all occupations can be automated entirely using demonstrated technology, about 60% of all occupations have at least 30% of worker made activities, that would be automated. More occupation will change to be automated. They also indicated for business performance benefits of automation are relatively clear, but the issues are more complicated by policy making to attract foreign investors. Beyond technical feasibility, the cost of technology, competition labor will include skills and supply and demand dynamics, performance benefits and beyond labor cost savings and social and regulatory acceptance will affect the automation. Their predictions suggest that half of today work activities could be automated by 2055 year, but this could happen 10 to 20 years earlier or latter depending on the various factors in addition to their wider economic condition.

Some scientists suggest (AI) technology is finally starting to deliver real-life business benefits. Computer power is growing significantly , algorithms are becoming more sophisticated and perhaps most important of all, the world is generating vast quantities of the fuel that powers (AI) technology data billions of gigabytes of it every day. Also, online firms are digital natives, such as Google online search service company is investing on (AI) technology. For new though most of the news if coming from the suppliers of (AI) technologies. And many new users are only in the experimental phase. Few products are on the market or are likely to arrive these soon to drive immediate and widespread adoption. As a result, analysts believe (AI) technology's potential will give true economic benefit in the future. (AI) industry will introduce to suppliers and users to raise economic potential of (AI) technology.

In the future, (AI) technology systems can solve business problems. Some scientists categorized those into five technology systems that are key areas of (AI) technology development: robotics and autonomous vehicles, computer vision language virtual agents and machine learning , which is based on algorithms that learn from data without replying on rules-based programming in order to draw conclusions or direct an action.
Such as computer vision and language includes natural language processing, analytics, speech recognition technology, some are about learning from information, such as about machine learning and others are related to acting on information, such as robotics, autonomous vehicles and virtual agents, which are computer programs that can converse with humans. Machine learning and a subfield called deep learning are artificial intelligence applications.

Can AI excite online leisure industry development?
Artificial intelligence (AI) is a term first defined in 1956 year. It is a branch of computer science that aims to create intelligent machines that work and react like humans. In contrast today, 60 years later, (AI) is characterized by a number of applications, including computers playing games against humans and understanding human languages, virtual personal assistants, and robotics which involve computers seeing , hearing and reacting to sensory stimuli. In the future, technologists predict for (AI) technology ranging from (AI) being used as a tool to aid relatively simple processes for robots with human like mental capabilities, who expect (AI) technology can emulate human performance by learning, coming to mind its own conclusions, understanding complex content, engaging in dialog with people, enhancing human cognitive performance or replacing humans in executing both routine and non-routine tasks. In existing industry, (AI) technology is used , such as targeted advertising and virtual used personal assistant as well as the (AI) technology that my exist in the future, such as robots with human vehicle processing capabilities.
The range of (AI) technology's progress in the future will determine the economic impact future of (AI) technology on the global economy with more limited advances and applications (i.e. weak (AI) only) corresponding to more limited economic impacts and more substantial progress, i.e. strong (AI) technology is corresponding to more significant economic impact.
(AI) technology learning that automates analytical model, including

predicting cause-and-effect relationship from biological data, identifying new drugs, self-driving cars and protecting against fraud etc. functions. Also (AI) learning can improve natural language processing that allows computers to continue to better analyze, understand and generate language to interface with human using the natural human language, virtual personal assistant, helps users by providing scheduling appointment, reminds organizing personal finance and finding providers of various services, machine vision allows (AI) machine man to identify object, scenes and activities in detect pedestrians and bicyclists.

We expect the economic effects of (AI) technology to include both direct GDP growth from sectors that develop or manufacture (AI) technology and indirect GDP growth through increased productivity in existing sectors that employ some from of (AI) technology. If (AI) producing sectors could grow, then it could lead to increase revenues and employment of (AI) technological professionals within these existing firms as well as the potential creation of entirely new economic activities to any countries' societies productivity improvement in existing sectors could be realized through faster and move efficient processes and decision making as well as increased (AI) technological knowledge and access to information available in societies easily.

In the future, if (AI) technology is an increasingly critical component of more products, it will become an integral part of necessary products of many people's lives. The extent of (AI)'s economy effort is also likely to vary from region to region, thought variation may be more dependent on the predominate economic activity of a region and the (AI) ability can influence economic activity, rather then the economic or developmental status of the regions. (AI) technology can move accessibility and can use source development to do international business between one country and another country.

So (AI) technology has the potential to give benefits to different income chooses and to bring significant gains to both developed and developing countries. For agricultural technology, (AI) has the potential to optimize food production around the world by analyzing agricultural regions and identifying what is necessary to improve crop yield. In total, (AI) technology gives greater economic impact to any countries agricultural regions if which implemented (AI) technology to grow crop , fruit etc. food production in the farms.

Investment in (AI) technology is such as capital investment to any countries' public or private enterprises. So, it will have large economic impact to the future . If the (AI) technology is reasonable invested to the different needs aspect by the public or private enterprises in the country. Then, it will have good economic impact to the country in the future. However, when (AI) technology is likely to affect both the productivity and employment components of economic growth in many sectors. Significant public debate has focused on projections of (AI)'s effect on the labor force. However, for instance, some researchers have argued that the rise of (AI) technology and automation will led to significant unemployment as capital is substituted for the low skillful labor. So, they point to the concern that the increasing sophistication of (AI) technology may balance skilled and semi-skilled workers and the reduce the size of the middle class. However, this is not a new argument, due to (AI) technology negatively affecting the labor force and leading to mass unemployment. Because the (AI) technology is the substitution of machinery for human labor. Although, employment in certain industries, has been reduced in the past due to technological advancement. For long term, the labor market has adapted to the introduction of new technology, giving rise to new jobs in new areas. (AI) technology may also be accomplished without a reduction to total employment in the long-term to some Asia countries, such as Hong Kong and Japan. Because Hong Kong and Japan many low skilled labor, e.g. security, cleaner who complaint that employers need them to work long time hours. (abnormal working hours) e.g. one day 12 to 15 working hour per day. Hence, if (AI) machine means invention technology success. Security or cleaning job can be worked from (AI) machine man in some hours every day in order to reduce the long time working hours cleaners or security workers, e.g. one (AI) machine man works 4 hours for cleaning or security job, one day as well as another cleaner or security labor only needs to work 8 hours one day. So total security or cleaning employers can employ 12 hours machine cleaners or security workers and human cleaners or security workers in one day. For long term benefit, Hong Kong or Japan every security or cleaning worker does not need to work 12 hours minimum working hours one day. They won't feel tried and bore and without private with whose families, so who will accept to do these cleaning or security jobs, even they can raise work efficient and performance when who feel happy and health.

So, (AI) technology of machine man invention can raise low skillful labor efficiency and it can help them to avoid abnormal working hours demand in some busy work life countries, such as Hong Kong and Japan. Before, one Japan female labor feel unhappy to work, due to who often needs to work abnormal working hours for her employer and who has less sleeping and without any private time to enjoy her life with her families every day. So this abnormal working hours factor causes her to do commit suicide behavior, then she is die unlucky. So (AI) technology of machine man invention ought avoid abnormal working hours demand for employer in any countries in the future.

The most important occurrence to any employers, some researchers had attempted to do one experiment to find that private research and development , venture capital and public research and development investment all have strong net effect or economic growth with venture capital funding further having the strongest such effect from (AI) technology. The researchers hypothesize the venture capital investment contributes to economic growth through (AI) technology innovation and by the capacity of an economy to use existing (AI) technology knowledge to increase productivity. They predict the impacts of venture capital, business-research and development and public research and development can raise multi factor productivity from (AI) technology introduction.

Can (AI) technology influence the economic development to developing countries? The developing regions of the world contain most of natural resources. If one day, (AI) technology has invent one kind of machine man which can assist any gas or oil workers to seek any new oil/gas natural resource locations easily. I believe that (AI) technology can help these natural resource exploitation countries will gain economic benefit more easily. So, (AI) driven technology can be used to change to create any new opportunities to address poor management or resources and improve human well being, such as Africa Latin America and India can use (AI) technology machine man to seek any oil/gas natural resource countries exploitation activities to attempt to gain much economic benefits.

- Why will (AI) technology grow economic
development ?

Nowadays, increases in capital and labor are no longer driving the levels of economic growth, such as (AI) technology. The ability of increase in capital investment and in labor of traditional drivers of production, have no longer to be enjoyed in most developed economies ,e.g. developed country, US, UK . However, artificial intelligence has the potential to overcome the physical limitation of capital and labor to avoid missing out on this opportunity.

So, policy makers and business leaders must prepare for and work toward a future with artificial intelligence. They must do with the idea that (AI) is another simply method to enhance productivity method . Rather they must see (AI) as the tool that can transform thinking about how growth is created.

Economists have always thought of new technologies are as driving growth their ability to enhancing. It can replace labor and capital factor of production. So, it brings this question: What is the factor of production (AI) technology characteristics. They key factor is to see (AI) technology as a capital-labor .

(AI) can replicate labor activities at much greater scale and speed, and to even perform some tasks began the capabilities of human. For example, by using virtual assistants , 1000 legal documents can be reviewed in a matter of days instead of taking three people six moths to complete. Some (AI) technology may be one kind of factor of production in the future. For another example, people will work in workplace digitalization environment. So, in the future, working environment and information management are automated. Such as Konica camera sale company will use workplace digitalization. So , (AI) technology can provide workplace digitalization in order to raise productivity efficiency. (AI) technology will be one kind of production which is replaced by workplace digitalization and it will grow any organization productivity efficiently. Then, (AI) technology will assist overall social economy growth , due to productivity is raised and products can be produced in short time to prepare to sell in consumption market. So, time will be shortened to increase GDP growth fast for the development of (AI) technology countries.

● How can (AI) technology impact to global economic and social and psychological changes?

What will be the development of (AI) technology and predictions concerning the future evolution? The computers and robots will develop conscious, intelligent and minds into humans, enhancing psychological and behavioral abilities and allowing for direct communication with (AI) minds. (AI) technology will be impacted human life by (AI) technology information communicative and environmental influence. A " world brain" and " world mind", this psychological system will be enhanced and enriched the capacities of both individual and collective cognition by (AI) technology of service industries.

(AI) technology with influence these human needs of service industries changes, such as , biological science, finance, entertainment, business, biological science, transportation, communication military etc. The personal computer evolution, the internet and the world wide web which exploded on the scene, linking business, homes, schools, social organizations which were a completely unpredicted phenomenon to influence human life. Kurzweil (1999) predicts that by 2029 year, most human communication will be with machines. According to Person, by 2100 year, there will be human machine convergence.

How can (AI) technology influence environmental protection to make benefits to farming economic growth? (AI) technology can be applied to predict how to solve environmental pollution challenge to avoid to damage any crop or vegetable or rice or fruit etc. food growth. Because environmental experts can gather global environmental pollution data from an environmental database to build a perform a systematic analysis from (AI) technology. The first step is this broad analysis can include understanding, statistical and data gathering techniques to obtain the relevant data, the correlation among the variables involved, and a list of possible models. The next step is to select a set of methods and models that cover all kinds of knowledge and functionalities needed for the decision making process. Once the models are selected, they must be fully implemented by means of machine learning , data mining, statistical or numerical technique. After that, those models must be integrated to build the whole EDSS. The EDSS must be tested to check its performance, accuracy, usefulness and reliability, both from the user's and (AI) technology/computer scientist's point of view. If these is any wrong feature in any development stage, such as model's integration, models' implementation, selection of models, database, problem analysis etc. the developers must come back in the update th required components. When the evaluation phase is all right, the EDSS is ready to be applied to the environment. The great contribution of artificial intelligence to EDSS the integration of several methods complementing the classical statistical models/simulation , statistical analysis, linear models, etc. and numerical models (control algorithms, optimization techniques etc.) .

This cooperation makes the resulting systems more reliable and powerful in coping with real world environment systems. Date interpretation has been a principal area of research in (AI) technology since the very beginning.

The most demanding problem in the environmental assessment context. Knowledge representation permits the definition of the different types of data that the existing methods adapt to the process. There is also a lot of work to clean, repair and transform the huge available quantities of raw data. Apart from this, the availability of meta-information or background knowledge is required to guide the process. Data mining is multi-disciplinary: It covers expert systems, data based technology, statistics, data visualization and unsupervised machine learning. These techniques operate at the level of data and background information, where numerous and often incompatible new commensurate pieces of information from disparate sources have to be brought together (K, Fedra, 1994).

So, it seems that in the future, (AI) technology with the increasing maturity in particular those related to knowledge and engineering, new dimensions can be assisted to users in environmental decision making are available. For example, many environmental systems are characterized both by incomplete models and by limited data. Hence, in the future, (AI) technology will be applied to predict climate change to reduce crop or fruit etc. food agriculture challenge by climate change bad influence.

- Will (AI) technology influence new kinds of leisure industry development ?

To understand how the manufacturing business must adapt to prosper in the technology, we need to understand how (AI) technology will change us to shape our daily habits to satisfy our expectation of products to how we shop and even the immediate of the entire process. For example, taxi services are in the crosshairs as on demand transportation services like, available of the touch of a smart phone button expand. In fact, Yellow lab, US country , san Francisco city's largest taxi company is filing for bankruptcy as the industry starts to change faster than almost anyone expected. However, at this point, its more than an app that is changing, some our taxi passengers renting taxi transportation to catch consumption behavior.

(AI) technology will influence digital economy for taxi passenger's individual customer experience, offering a growing renting taxi to catch of service and feedback opportunities when any one taxi passenger who chooses to use mobile phone app online tool to prepaid to rent any taxi more easily.

Also in the long term, (AI) technology can influence vehicles drive themselves of behavior. Already, companies like Google and GM are working on projects to bring fleets of autonomous vehicles to cities at the path of a button.

Moreover, this on-demand service model is beginning to appear across a much broader range of markets. For example , Amazon company is investing in its own fleet of trucks, planes and even drone at the same time as it pushes for same-day delivery of products. As some point, vehicles will be autonomous too. So, it seems that (AI) technique will influence any transportations choose to use digital autonomous driving technology in the future . For Amazon company case, it is not stopping of logistics. It is also aiming to automatically manage the supply of consumer home products with its recently launched Amazon replenishment service, Dash. Dash is a digital service that enables that connected derive to automatically order physical products from Amazon when supplies are running low. So, it seems (AI) technology will be applied to logistic function by digital technology method introduction in the future.

Hence autonomous vehicles will optimize industry supply chains and logistics operations through increased efficiency and flexibility. In fact, fully automated and lean supply chains will keep reduce load sizes and inventory by leveraging smart distribution technologies and smaller autonomous vehicles by machine man assistance. If Amazon continues to grow market share for online sales by reducing effort required by the consumer to place an order, when also contributing the almost immediate delivery of products to the doorstep. So, it will further fuel the trend toward on-demand derive. As Amazon company fuels the on-demand economy, consumers will expect immediacy in more parts of the digital economy. On top of speed, consumers increasing expect more personalization options.

So, (AI) technology will influence digital manufacturing, such as Amazon publishing to monitor every aspect of every process in real -time and communicating to self-optimized deep learning robotics, new methods of high volume and high customization will become possible. Then, as products merge into product platforms and even services, manufacturers have the opportunity to provide components and platforms used by smaller players. So, (AI) technology will influence manufacturing industry to choose automated SMI lines, robots installed, automation engineers.

Another future (AI) technology development can be applied to space science aspect, such as Automation engineering space in manufacturing process to achieve digital manufacturing benefits to any businesses in the future. Such as reducing cost, shortening manufacturing time, raising efficiency, shortening delivery products to client individual time. How can artificial intelligence give the need and advanced fast and evaluation methods benefits for space exploration? When US NASA (space exploration organization) achieves any space exploration missions, it will answer this question:

When is it useful to have a machine use (AI) technology to achieve a decision? After all, after millions of years of space exploration and rough 10,000 years of civilization, humans are usually quite good at making decisions in complex uncertain environments. Through, Johns Hoplains University's Applied Physical Lab. Research in (AI) technology enabled systems, which has identified three general use cases for (AI) technology to explore space mission:

First, for some tasks (AI) technology is more cost effectiveness than human. Second, (AI) technology is better suited than humans at solving some, but not all problems. Third, (AI) technology allows NASA organization's space exploration mission to develop machines that ate capable of responding faster than when a human is in the decision loop (D. Scheidt, 2012, A. Castano et. al. 2008).

So, the use of (AI) technology to enable science by observing the pace of rapidly evolving phenomena was demonstrated. It is more effectively coordinating and (AI) technology utilizing to earn economic benefits to use for space exploration mission.

However, (AI) technology also have current risk for space exploration. Today (AI) technology is immature and requires further development to reach its potential. For instance, the (AI) technology algorithms that detected the dust derive could not have identified whether the Martain weather represented a threat to the cover. Also it can not yet use instrument input to determine what, where and how to autonomously make the next space science measurement. An equally important factor limiting (AI)'s deployment is that lacks the methodology and technology to effectively test (AI) technology. So, the challenge will testing (AI) enabled system is how (AI) performance can be measured. It would be NASA organization's difficulty to find (AI) technology to develop to carry on researching any space exploration missions in the future. However, (AI) technology will be a good economic benefit choice for space exploration mission in the future.

● How can (AI) technology influence to global health care economy development?

According to (AI) lecturer analysis, when combined key clinical health (AI) application can potentially create $150 billion in annual savings for the US healthcare economy by 2026 year. (AI) technology is re-winning modern conception of healthcare delivery. It enables machines to sense, comprehend, act and learn. So which can perform administrative and clinical healthcare functions (Accenture, 2017).

It will help health care service organizations to reduce health care cost, will improve and raise service quality and access. So, (AI) health market size will be predicted growth. (AI) applications in health care include robot-assisted surgery, virtual nursing assistant, administrative workflow assistant, fraud detection, error reduction connected machines, clinical trial participant identifier, preliminary diagnosis, automated image diagnosis and cybersecurity.

What kind of benefits (AI) technology can contribute to healthcare service? (AI) technology can deliver what many health care organizations need, such as financial and operational of labor costs, digital expectations from patient consumers how to use (AI) technology to solve interoperability challenges in any healthcare organizations. Also (AI) technology can be applied to wellness an d lifestyle management, diagnostics, delivers financially but also way of organizational and workflow improvement. So, (AI) technology will be continue to become most prevalent and adoption to healthcare organizations , which must need to enhance structure to be position to take full advantages of new (AI) technological capabilities. (AI) technology can change the nature of work and employment is rapidly changing to make the best use of both humans and (AI) talent in healthcare industry in the future. For example, (AI) technology offers a way to fill in gaps and the rising labor shortage in healthcare. According to Accenture analysis, the physicians shortage is increasing. However, (AI) technology will manufacture healthcare machine men to replace physicians in future one day(2017). Hence, (AI) technology will be invented to raise health care service staffs work efficiency and performance in any hospitals or clinics in the future.

In conclusion, (AI) technology will raise efficiency for any service or manufacturing industries in the future, although, it is possible that it will also rise low skillful workers unemployment numbers. But, the most important influence to human technological innovation will be risen and it will influence human life will be changed to be better, e.g. self drive cars, health care physician machine men, machine man cleaners etc. intelligent

machine men will be manufactured to serve for our daily life. Furthermore, (AI) technological products will influence countries trading, some low technological development countries manufacturing businessmen can choose to buy any (AI) products to raise whose productivity and efficiency and reducing cost to achieve economic cost saving result. Also, GDP of trading growth income will increase to the (AI) products sale countries. Hence, it will be beneficial to economic development to both developed and developing countries both in the future as well as (AI) scientists time and money spending will be valued to continue to invest (AI) technology development for human life and economy benefits for long term.
In conclusion (AI) technology will raise macro economy growth and it can create many (AI) jobs , but it also raise the low level technological worker unemployment change. In the future, (AI) technology can be applied to digital technology to attempt to invent any new undiscovered (AI) and digital technology. So, it needs any scientists to continue to research how digital and (AI) technology can be mixed to satisfy human's future undiscovered needs.

Must Developed And Developing Countries Need Artificial Intelligent To Replace Human Job

Must developed and developing countries need artificial intelligent development? If one developed country, e.g. US, UK , Japan , Singapore it does not continue to develop artificial intelligence, robotic, then what disadvantges or weaknesses , it will encounter to compare when it chooses to continue to develop this artificial intelligent technology in society. If one developing country, e.g. China, Korea, Taiwan, it does not continue to develop artificial intelligence, robotic, then what disadantages or weaknesses, it will also encounter to compare when it chooses to continue to develop this artificial intelligent technology in in society. I shall explan the reasons why the results may cause to either the developed country, or the developing country as below:

- How AI help developing countries to communication and agriculture and learning and medical delivery development

Why can AI help developing countries ? Drones that pick inaccessible crops and mobile phones that give medical advice are two of the ways AI can transform life in the developing world. Artificial intelligence (AI) may improve the lives of the world's poor, the technology needed to revolutionise inefficient, ineffective food and healthcare systems in developing countries is well. For example, in low-income areas, agriculture and healthcare are two critical ecosystems that we can apply AI to immediately; this is not the far future, or even in five years.

Artificial intelligence (AI) has seeped into the daily lives of people in the developed world. From virtual assistants to recommendation engines, AI is in the news, our homes and offices. There is a lot of potential in terms of AI usage, especially in humanitarian areas. The impact could have a multiplier effect in developing countries, where resources are limited.

Emergency Response to developing countries' earthquake natural damage suddence occurrence predicting

AI and machine learning are still finding importance in emerging markets, but certain applications have emerged and are now widely used. For instance, predictive models for disaster relief enable first responders to automatically analyze large-scale behavior and movement through multiple sources of data including social media platforms, web forums, news sources, etc. Based on collected data, responders can scale reconstruction efforts and distribute supplies in a timely manner.

Why and how AI can assist farmers to predict when the earthquake occurs suddenly in order to avoid or reduce the natural damage to their agriculture productive number loss. For example, In 2015, when a major earthquake hit Nepal, more than 8 million people were affected. During the aftermath, drones were used to map and assess the destruction and speed up the rescue mission. The town of Sankhu, situated about 20 kilometers northeast of Kathmandu, was among the highly affected locations. In May 2018, my company Fusemachines and GeoSpatial Systems partnered with Sankhu's city officials to use drones and artificial intelligence in an effort to automatically estimate the reconstruction need. After processing data accumulated from a drone-powered aerial mapping of the region, the team fed this data to advanced machine learning algorithms. Combining drone imagery, digital mapping and machine learning, the team configured region modeling and infrastructure development with higher accuracy. Another organization known as One Concern, a California-based startup, has created a predictive AI program called Seismic Concern to accurately predict seism and is also working on solutions for wildfires, floods and hurricanes.

Smart AI Agriculture

Another application of AI in developing countries is smart agriculture. Farmers monitor crops more effectively and make better predictions on planting, weeding and harvesting using AI tools. It can also be used to analyze one plant at a time and add pesticides only to infected plants and trees instead of spraying pesticides across large swaths of crops. One California-based tech company is an example of this use of AI. So, the developing countries farmers in rural parts of India are also using AI to increase yields through better access to information about the farming season than they would normally have. Technology-enabled process automation offers the agribusiness industry the chance for remarkable growth -- not only in developed countries but around the world. There's a unique opportunity to increase yields, cut down labor costs and improve people's health.

Medicine Delivery to developing countries' patients urgent need

Companies are also leveraging AI to improve access to health care in some of the most remote areas of the world. In Rwanda, for example, Zipline is using drones to deliver medical supplies and blood to hospitals and clinics that are difficult to access by car. This has dramatically impacted people living in remote parts of the country because they are able to get medical help when needed. The drone system in Rwanda has also helped reduce waste of blood by 95%, as noted by Zipline. One Concern has created an AI program called Seismic Concern that accurately predicts seismic events and is also working on solutions for floods, wildfires and hurricanes. The medical field may actually benefit the most from emerging technologies in developing countries.

Assistance to reduce teaching work workload or psychological pressure to teachers in developing countries' schools

Another vital area benefiting from innovative technologies like AI is education. Advanced technologies can enhance how we learn, teach and perform tasks. In most developing countries, schools lack experienced teachers and resources to enhance students' knowledge. As a result, many students still have to walk long distances to get to the nearest school, which has created education gaps, especially in rural areas. AI tools such as personalized learning assistants can simplify learning by making tutoring services and learning materials accessible to all students, wherever they are. Machines can be automated to help students learn basic concepts without a tutor, which companies like Carnegie Learning are working on. This would allow students to learn at any time from anywhere. With AI, education is made easy and accessible to more people.

The initial usage of AI in developing countries has been at a micro level --

solving small, specific problems in a defined industry. As machine learning advances and there is a higher utilization of AI, we will see more complex issues being targeted and resolved. When duly adopted, AI can positively impact future developing countries people everyday lives not just in disaster intervention, education, health care and agriculture but can also help in mitigating poverty, malnutrition and pollution. Especially, in developing nations, to leverage AI's true potential and create a snowball effect. Startups are defining a holistic and humanitarian approach to building more sophisticated, AI-ready societies. Stakeholders in the AI landscape should understand the strengths and nuances of the developing world as well as the limitations of AI and create localized solutions and applications.

Why does smart phone help developing countries communication ?
Internet Seen as Positive Influence on Education but Negative on Morality in Emerging and Developing Nations. Internet access differs substantially across the 32 emerging and developing countries polled, with the lowest rates of internet use in South Asian and sub-Saharan African nations. Within countries, computer owners, young people, the well-educated, the wealthy and those with English language ability are much more likely to access the internet than their counterparts. To access the internet, people increasingly use smartphones rather than more cumbersome fixed landline connections and computers. Around the world, both smartphones and basic-feature phones alike are used for sending messages and taking pictures.

In fact, many developing countries young people, students are popular to use smart phones for internet usage aim, instead of communication. Moreover, many developing countries working people are also popular to use smart phones for any working usage in their working time , even non working time any time. So, smart phones (AI) phones will be important communication or leisure tools to developing countries people in the future. Unless, it is one day, scientists can develop another new communication tool to replace smart phones. So, artificial intelligence will be important to influence developing countries people , how to improve or bring positive learning attitudes to students in their daily learnnng lifes. as well as how to raise developing countries people, how to raise working people efficiency or improve performace in their daily working lifes. So, AI may bring positive learning or working attitudes to developing countries working people and students both.

The Positive Impact of Mass Media in Developing Countries

Radio, newspapers, television, Internet, social media, etc., all of these are forms of mass media. Each of these outlets has the capability of bringing information to thousands of people with one device. While in some communities it is easy to take advantage of these communication outlets such as television and Internet access, not everyone has access to such outlets. Radio is one of the most common forms of mass media in developing countries because it's affordable and uses less electricity than many other forms of mass media, but only approximately 75 percent of people in developing countries have access to a radio, and roughly 77 percent of people in rural areas have access to electricity.

For developing countries that have implemented forms of mass media in their communities, there have been numerous positive outcomes are influenced to impact developing countries mass media by artificial intelligence as below:

When AI is participated to developing countries mass media, it can influence any radio, television audiences raise more attention to each other through social media platforms such as Facebook and Twitter and create, organize and initiate street protests and campaigns. Furthermore, having access to social media in developing countries, people are able to connect to those that they usually wouldn't have the chance to talk to. Moreover, AI Provides educational opportunities- In many countries, the division between local and national languages as well as issues of literacy can make communication difficult. With the use of mass media, a bridge can be built between these two gaps. In India, there is a radio station that provides information in local languages and respects local culture and traditions. One of the main ways is to create public awareness of what is going on with businesses and government officials. The media plays an important role in giving people the opportunity to act against injustice, oppression and misdeeds that they otherwise wouldn't know about. Information on available healthcare, a mass radio broadcast was sent out encouraging parents to seek treatment at local healthcare facilities for their sick children. With this mass outreach on healthcare, the encouragement of people to take their children to healthcare facilities saved thousands of lives. This easy way of encouraging others and bringing awareness about certain diseases was made possible through a simple radio broadcast. Finally, when AI is particiapted to media, it may bring many social issues to life that otherwise would remain unknown to many people. In developing countries and communities like Burkina Faso, when the radio broadcast was released about malaria, diarrhea and pneumonia, people were educated and moved to action and knew to take their children to healthcare facilities for preventative care. As it is seen, having access to different media outlets is vital for those in developing countries. Here are three ways that those in developing countries can implement mass media to help their people and communities.

When AI is participated to any internet radio or internet newspaper mass online listening or reading channel. It can provide online radios or newspapers in public places- By providing online radios and newspapers in public areas it gives community members to access news, information and emergency warnings. Even though radios can be on the cheaper side, there are still many people that can't afford to have a radio in their home. By providing one in a local place, not only would it better educate the community members but also it will bring the community together. So, it can make media outlets a two-way platform- Creating a two-way platform between the community and those who are behind the radio stations, newspapers or broadcasts makes the community feel involved and that their voices are being heard. An organization called Soul City in sub-Saharan Africa is showing how well two-way platforms work by engaging their listeners and having them contribute thoughts and ideas about complex issues. Because developing countries radio listening audiences or newspaper readers are popular to accept computer online radio listening channel or online newspaper reading channel to replace traditional paper newspapers or radio machines. So, AI may raise their listening news or reading news leisure feeling from online mass media channel in the future.

In conclusion, when developed countries continue to develop AI, it may bring positive advantages to bring raising productivies, or efficiencies, but it may also raise unemployment ratio to any low skill or low knowledge jobs in ther societies. However, human future society will need to change to be better to raise our living standard. But AI is one kind the best choice tool to achieve this aim in our future, so I agree developed countries continue to develop or research AI to be the super -human machine.

Reference

A. Castano et. al. " Automatic detection of dust devils and clouds at Mars" Machine vision and applications, Oct. 2008, vol. 19, no 5-6, pp. 467-482.

Accenture, " Why artificial intelligence is the future of growth"(2017) <http://www.accenture.com/us-en/insight-a rtificial-intelligence-future-growth>.

D. Schedidt , Unmanned Air Vehicle Command And Control, Handbook Of Unmanned Air Vehicles, Springer-Verlag, 2014. Facebook (AI) Research Available at https://research.facebook.com/ai, research at google, machine intelligence available at http://research.google.com/pubs/machineintellige nce.html; micro soft research-machine learning and artificial intelligence available at http://research.microsoft.com/en-us/research- areas/machine-learning-ai.aspx.

International Federation Of Robotics, 2016. IFR press release world robotics report. IFR, org . 29 Sept. Accessed Feb. 01, 2017. http://www.ifr.org/news/ifr-press-release/world-robitics report -2016-8321.

K, Fedra , "GIS and environmental modelling" in environmental modelling with GIS, edited by M.F. Goodchild.B.O. Parks and L.T. Steyaert, Oxford University press, pp. 35-50, 1994.

Keynes, J.M. (1933). Economic possibilities for our grandchildren (1930). Essays in persuasion, pp.358-73.

Mckinsey & Company (2013, May). Disruptive technologies: Advices that will transform life, business and the global economy , USA.

Ministry of economy, trade and industry, Japan, 2015, Japan's robot strategy. Ministry of economy, trade and industry.

Ray Kurzweil , The age of spiritual machines (1999) is cited numerously through this chapter: Kurzweilai.net http://www.kurzweilai.net

Rich, Elaine & Knight, Kevin, Artificial Intelligence Second Edition, 1991, New York; Mc-Graw-Hill.

www.ingramcontent.com/pod-product-compliance
Ingram Content Group UK Ltd.
Pitfield, Milton Keynes, MK11 3LW, UK
UKHW041829200726
13854UKWH00002BA/891